COMPUTERS AND LITERATURE

COMPUTERS AND LITERATURE
a practical guide

B. H. Rudall

and

T. N. Corns

University College of North Wales

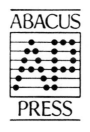

Cambridge, Mass
&
Tunbridge Wells, Kent

First published in 1987

ABACUS PRESS
Abacus House
Tunbridge Wells
Kent, TN4 0HU, UK

and

ABACUS PRESS
PBS, PO BOX 643
Cambridge
Mass 02139, USA

Printed in Great Britain by
Antony Rowe Ltd., Chippenham

Library of Congress Cataloging-in-Publication Data
Rudall, B. H. (Brian H.)
  Computers and literature.

Bibliography: p.
  Includes index.
  1. Literature—Research—Data processing. I. Corns,
Thomas N. II. Title.
PN73.R83 1985 802′.8′5 85-13370

British Library Cataloguing in Publication Data
Rudall, B.H.
  Computers and literature: a practical guide.
  1. Literature—Study and teaching—Data
processing
  I. Title II. Corns, Thomas N.
  802′.8′54 PN61

ISBN 0-85626-340-0

It were good . . . that men in their
innovations would follow the example
of time itself; which indeed innovateth
greatly, but quietly, and by degrees scarce
to be perceived.

Francis Bacon, 'Of Innovations', *Essays* (1625)

PREFACE

This book is the product of collaboration between practitioners of computing and literary criticism. We address ourselves to colleagues and students in both disciplines and to any who wish to explore the strengths and limitations of the computer in the study of complex human communication. Our target readership is diverse in background, objectives and expertise. Among literary researchers, there are many who have been impressed by reported results in established areas of computer investigation, such as concordance generation or investigation into questions of authorship, and who need to know how they can start such projects. Generally, among workers in the humanities, reservations and, perhaps, anxieties about computers have, very properly, been eroded as word-processors and home computers have become commonplace. This book has much to say to those who wish to adapt to literary studies the power and convenience of data-management and word-processing procedures. We address ourselves, too, to computer scientists, to explain the sort of problems involved in literary research and to suggest the strategies most appropriate to their resolution. Students of computing will be able to understand better the sorts of projects where the computer can help in the study of literature, and thus be better prepared to meet those challenges. Moreover, our approach is both practical and theoretical and will, we hope, suggest many areas for the development of computer applications. We fully discuss the role of microcomputers and a whole chapter is devoted to them. Those with a more general enthusiasm for computing may well be interested in the formulation and resolution of problems, some of which may be adapted and worked through using the modest resources of a small personal computer.

Readers with a good background in computing may find it appropriate to read chapters two to five more cursorily. Chapter two is designed primarily for the complete novice in using computers, while chapter three orientates the more experienced computer user towards literary applications and chapter four deals with programming languages which are appropriate to this area.

Canonical investigators may be curious to know that chapters one, six, seven, eight, nine and ten were, for the most part written by T. N. Corns, the rest by B. H. Rudall, but both, of course, accept responsibility for the whole. We should like to acknowledge our gratitude to SPSS Inc., for permission to describe the SPSS suite of programs and reproduce output generated by it; to IBM Ltd., for permission to reproduce a photograph of a word-processing computer; to the staff of the Eighteenth-Century Short-Title Catalogue, and especially M. J. Crump, for full and kindly response to our queries; to the Computing Laboratory and the Library of the University College of North Wales, and especially to M. Yamaguchi, for her assistance in accessing ESTC; to Prof. G. Guffey, for a number of suggestions; to J. Borland, for assistance with the design of the project described in chapter eight; to P. Kitchen, for typing sections of the manuscript; and to Tony Griffiths, for the technical drawing he produced.

November 1986

Brian H. Rudall

Thomas N. Corns

CONTENTS

1. INTRODUCTION

Formerly, the use of computers in the study of literature occasioned hostility and controversy. To scientists and technologists, the major users of academic computing facilities, the intrusion of the student of literature often seemed inappropriate, the dissipation of scarce resources. From computer scientists themselves, the response was sometimes uncertain, as the literary neophyte struggled to articulate his problems and requirements in a terminology he but tenuously grasped, though, of course, we must at once concede that some of the pioneers were accomplished programmers. Similarly, the literary traditionalist felt an instinctive antipathy to the reduction of complex verbal structures to schemes and categories tractable to machine manipulation. To them, the activites of early workers in the field seemed rebarbative and repellent, a mechanical operation of the spirit, performed by people incapable of sensitivity.

That was some time ago. The first concordance work dates from 1949; the first major stylistic studies and the first attempts at author identification from the early 1960s [1]. Attitudes have mellowed and become more rational and non-numeric computer applications are now much commoner. Critics and scholars working in the literary field can now see much more readily that computer applications, far from invalidating other approaches, either perform familiar tasks much less laboriously and more precisely, or else disclose techniques for the realisation of traditional ambitions and objectives which could not have been accomplished simply using traditional means.

Some computer applications in literary studies remain, in the vital and constructive sense, controversial. Much of the work in progress is experimental, attempting to use the computer to resolve problems of the literary text which have not previously been treated using such techniques. However, there is by now a considerable body of experience which the literary scholar can draw on. A number of procedures have been adopted, some of them many times, to answer problems and produce results that are of interest to critics and

1

scholars. We shall, in this book, primarily be concerning ourselves with describing techniques which have worked, showing how to use them, and suggesting ways in which their application can be extended. Our aims are practical and instructional, not experimental or radically innovative.

Currently, only a small proportion of literary studies are computer-based or computer-aided. Yet, we are confident, the time is ripe for a quite major transformation. Most academic researchers have access to resources which, though stretched and scarce, are nevertheless there for the using. The use of computers is progressively easier, even for those without a background in scientific training. The machines are, increasingly, 'user-friendly', prompting the inexperienced how to use them as they go along, and documentation, too, becomes simpler, closer to the kinds of language that humanists are familiar with. Rapid developments in computer technology may well very soon produce new generations of small computers, little bigger than a typewriter, cheap and powerful enough to offer unlimited processing power to all researchers. Present-day microcomputers, about which we have quite a lot to say later [2], are, of course, a major advance towards this prospect. Finally, and most important, we have now a corpus of experience and a core of proven techniques that the critic and scholar can draw on. There is no good reason why they should not regard the computer as much an everyday research tool as the box-file and the microfiche reader, the typewriter and the photocopier. No doubt, eventually, they will. If our book has a larger purpose, it is to accelerate that process.

Computers can perform some tasks much better than others. They are excellent for holding large quantities of information, for retrieving it, for sorting it, for performing calculations on it, and for identifying patterns within it. They are much worse at understanding the information, and they are quite incapable, of course, of responding aesthetically to it. These strengths and weaknesses shape the way in which they may be used in the study of literature.

What do we mean by 'information' in our context? It can take several basic forms. We may store within the computer raw, uncoded pieces of natural language, the texts which we are working on, inserted into the machine more or less as we find them. There are no major problems in using a machine to store *Hamlet* or *Paradise Lost* in such a way that it can produce a typed copy of them on request (and, as we shall see, do much else with them besides).

Again, we can insert into the machine the sort of information which, at present, we may customarily store in notebooks and ring-files; facts, ideas, quotations taken from other authors, and our own working notes, information to which, we feel, we may in future need to refer. Further, it is possible, when we come to assemble our research findings for publication, we may store the various drafts and versions in the machine too.

A second class of information is rather more structured than natural language. This is the sort of information that is held on forms or data-sheets or else perhaps on cards in a box-file (or possibly many box-files). A couple of examples may help. Researchers may be engaged in a piece of theatre history, collecting information about the play in performance. For every play in their study, they may wish to hold a record of its dates of performance and composition, the actors, directors, designers, producers, the publishing details, if any, the reviews the performances received, and so on. Or descriptive bibliographers may wish to record, for each publication they are describing, its author, its publisher or bookseller, its physical size, its collation, and various pieces of miscellaneous information, such as the location of extant copies or references to it in other bibliographies. In both cases, formerly, the information gathered would necessarily have been kept on cards or slips in a box-file of some sort and then laboriously sorted and cross-referenced, item by item, by the research worker. Now, however, so long as the records are structured into appropriate formats, they can, with facility, be inserted into and held by computers, as we shall see, with many distinct advantages. Even the booklists and working bibliographies which are inevitably a part of the scholarly work can usefully be stored and manipulated by machine.

Finally, we may feed the computer with encoded information. This is the most structured sort of information. Instead of recording details about our texts in natural language, the information is held, rather, in a (usually) numeric code. Such a procedure is often appropriate if the variables associated with our texts have been categorised in some way. For example, we may be interested in an author's sentence structure and have assigned all the words in all the sentences in the material under analysis to syntactical categories. It would be sensible to encode such categories to facilitate the storage and manipulation of the information. Such a code may attribute to every personal pronoun the value 07 or to every transitive verb the value 38, and so forth. Information about the syntactical distribution of any sentence could be reduced to a series of numbers. Let us take a sentence: 'We have, then, three organizations of myths and archetypal symbols in literature'[3]. This could be represented by the numeric series 07 38 12 02 01 11 01 17 02 01 03 01. We shall see later ways in which such procedures can be useful. Again, information about other aspects of the text may be reduced to numbers in rather different ways. For example, the circumstances of its production may be encoded. Supposing we are looking at, say, the work of the press in a particular period, for example, the English Civil War. We may categorise all the items that interest us in terms of date of publication, place of publication, whether it bears the name of its author, printer and bookseller, whether it has an illustration, its collation, and how long it is. Such information can easily be translated into a numeric code in which each item in our study will be represented by 9 numbers:

1. its number within our study
2. its date of publication
3. it place of publication (01 London 02 Oxford 03 other 04 not given)
4. does it have its author's name or initials? (01 yes 02 no)
5. does it have its printer's name or initials? (01 yes 02 no)
6. does it have its bookseller's name or initials? (01 yes 02 no)
7. is it illustrated? (01 yes 02 no)
8. collation (01 broadside or single sheet 02 folio 03 quarto 04 a smaller format)
9. length (in thousand of words)

Thus, the numeric series 103 1648 01 01 02 02 01 03 010.5 would mean that number 103 in our study (we can cross-reference our project and look up exactly what this item is in the sort of computer-readable file we were discussing above) was published in 1648, in London; it bears the name or initials of its author, but is unacknowledged by either printer or bookseller; it has an illustration, is printed in quarto, and is 10,500 words long. Of course, such a profile is tractable to almost infinite refinement, and the information may be analysed statistically.

A multiplicity of aspects have traditionally exercised and concerned the literary critic and scholar. Let us list some, and decide where best the machine may help. Our focus may be on the text as a linguistic structure, as a physical object, or else on the circumstances of its production. Every literary text, of course, consists of words. It is tractable to stylistic analysis at several linguistic levels. What vocabulary does it use? We may want to list all the words, to see if they are hard words, slang words, dialect words, technical words, words newly coined either from native resources or through borrowing and so forth. We may want to consider the vocabulary of one text in comparison with other texts by the same or different authors. There are other lexical perspectives. How is the vocabulary being used? Are the words 'collocated' or used together in surprising or abnormal ways? Does the text use the same words for the same concepts whenever they occur, or can we identify sets of synonyms? Perhaps we want to look at grammar. Accidence may be interesting. As the author forms plurals, inflects verbs, and so on, there may be aberration from known linguistic norms. Sentence structure is a traditional and obviously valid area of stylistic enquiry. Then there is the semantic level. What are the thematic emphases of a text? How are they developed or expounded? This may lead to consideration of narrative structure, how the story is related, and the relationship between the narrator and the author whose values we perceive behind the text. Phonological aspects, rhythm, rhyme, alliteration, are often objects of study. The textual critic wants to establish the relationship between the text under study and other versions of the same verbal structure, generally as a stage in construction of an authoritative version closest to what the author originally intended. The descriptive bibliographer describes the text as a physical object with extreme accuracy and chronicles the details of its production and perhaps of its survival. Much of his information will be used by historians of publishing. Author-identification is a narrowly circumscribed but not unimportant area of enquiry, enabling us to find possible authors for anonymous or pseudonymous texts or challenging traditional attributions. There are source studies, searching for echoes, allusions, borrowings. The list is, of course, far from exhaustive, nor can we anticipate developments in the discipline. It would be unwise totally to discount useful, perhaps ingenious, computer applications in any of these approaches. But it is imperative that we recognise at the earliest stage of any project the limitations as well as the strengths of the machine.

Computers are weak at understanding natural language. Computer science has developed special, synthetic languages, called 'programming languages', through which to communicate instructions to them. The natural language of the literary text poses enormous problems at the semantic level.

It is possible to equip the machine with a dictionary in which it can look up the words of a literary text and thus identify synonyms or assign words to syntactical categories. There are, however, grave difficulties. Consider the sentences:

> The President gives a lead to the country.
> The President puts a lead into his propelling pencil.

No machine could, with facility, be called on to recognise two distinct root words, 'lead' (OED sb.1, signification 3, 'Short for BLACK LEAD, graphite, plumbago. Only with reference to its use as a material for pencils'), and 'lead' (OED, sb.2, signification 1, 'The action of the verb LEAD 1; leading, direction, guidance'). If you ask the machine to read a text and note incidences of the theme of 'leadership', even if it is equipped with a full vocabulary of appropriate 'leadership' terms ('lead', 'guide', 'direct', 'inspire', and so on), it will produce strange results when it comes to process the second sentence. Natural languages also present interesting obstacles to the identification of syntactical structures. Take the sentences:

> President Reagan signs lead control legislation.
> Bangor signs lead export drives.

To the speaker of English the differences in syntactical distribution may be apparent. Plainly, in the former, 'signs' is a verb and 'lead' a noun; in the latter, 'signs' is a noun and 'lead' a verb. The proper nouns, 'President Reagan' and 'Bangor', really function quite differently. The former is the subject of the sentence, the latter is used attributively, qualifying the subject 'signs'. For a speaker of English to effect this parsing, he has to draw upon evidence outside the discrete verbal unit. He uses his knowledge of the larger world to resolve the problems. He knows that lead, the toxic element, is a likely subject for legislative control, just as he declines to identify President Reagan as a source for commercial signs. He cannot make sense of the former sentence if he assumes the structure

Noun Phrase		Verb Phrase
	Verb	Noun Phrase
<<President Reagan signs>>	<<lead>	<control legislation>>

Similarly, he cannot make sense of the latter were it to have structure

Noun Phrase		Verb Phrase
	Verb	Noun Phrase
<<Bangor>>	<<signs>	<lead export drive>>

The English speaker, however, would rest content with

Noun Phrase		Verb Phrase
	Verb	Noun Phrase
<<Pretty Polly signs>>	<<lead>	<revival in poster design>>.

The correct analysis, of course, draws on a familiarity with English cultural codes, about hosiery and graphic design, which is not contained within the discrete sentence. It is grossly improbable that any machine in the foreseeable future will be equipped to make the sort of decisions the English speaker makes with facility. Perhaps the classic illustration of the problem is the following pair of sentences:

I went to the store that sells horse shoes.
I went to the store that sells alligator shoes[4].

The speaker of English recognises the differences between the attributive nouns 'horse' and 'alligator' because he is familiar with the culture of which the English language is a part. He knows that some kinds of shoes are made from alligators and that other kinds of shoes are made for horses. The machine, of course, is much less able to make that kind of distinction, less able to understand what the sentences mean.

This may all sound rather pessimistic, and certainly it has important implications for computer applications in the study of literature. Plainly, the machine functions badly at the semantic level and at those parts of the syntactic level which invoke semantic interpretation. It is possible to call on the machine to find the thematic concerns of the text, though it may do it badly. Again, the parsing of sentences, which is the unavoidable first stage towards the analysis of sentence structure or the identification of syntactical preferences, draws, as we have seen, upon levels of linguistic and cultural activity with which the machine cannot successfully deal. There has been some progress in the development of parsing programs, and this remains an area of considerable research interest. However, the best systems still require a high level of human involvement in assigning items to their syntactical categories. It is worth observing that the examples we have just looked at are all from the sorts of language we would find in modern prose exposition. Imagine the difficulties the machine would encounter in coping with the complexities and abnormalities in the language of Shakespeare or Milton or the later Joyce.

SUMMARY

This chapter really makes two points. Firstly, the historical moment is ideal for initiating computer activity in the field of literary studies. Secondly, we must recognise the strengths and limitations of the machine. Computers are weak at operations that involve the understanding of natural language. They are most useful at manipulating natural language in ways that do not invoke the semantic level, and in processing structured and encoded information, and there, as we shall demonstrate, their role is potentially very important indeed.[5]

NOTES

1. Susan Hockey, *A Guide to Computer Applications in the Humanities*, London, 1980, pp. 15–17. This book admirably reviews much of the work already done in literary computing.
2. See below, Chapter 5
3. Extracted randomly from Northrop Frye, *The Anatomy of Criticism*, 1957: New York, 1968, p. 139. The application is suggested by L T Milic, *A Quantitative Approach to the Style of Jonathan Swift*, The Hague, 1967.
4. The example is taken from J Katz and J Fodor, 'The Structure of a Semantic Theory', *Language*, 39 1963, 170–210, discussed by Philip Hayes, 'Semantic Markers and Selectional Restrictions', in Eugene Charniak and Yorick Wilks (eds), *Computational Semantics; An Introduction to Artificial Intelligence and Natural Language Comprehension*, Fundamental studies in Computer Science No. 4, Amsterdam, 1976, pp. 41–54 (p. 50).
5. For a consideration of the connections between recent theories of criticism and computer applications, see Thomas N. Corns, 'Literary Theory and Computer Criticism: Current Problems and Future Prospects', in *Méthodes quantitatives et informatiques dans l'étude des textes,* edited by E. Brunet, Geneva, 1986, pp. 222–227.

2. INTRODUCING THE COMPUTER — THE CONCEPT OF THE 'LITERARY' COMPUTING MACHINE

This chapter is primarily addressed to those readers who know little about computing. Its purpose is to introduce them to the constituent parts of the machine and the way in which they function. It explains the fundamental operations of input and output, of storage, and processing. Those with some familiarity with the machine may confidently read cursorily or turn to chapter three.

THE 'LITERARY' COMPUTER

Any currently designed computer can be considered a 'literary' computer in as much as it will accept input information concerned with this field, process it and produce output information. Figure 2.1 is a schematic diagram of a typical computer architecture looked at from the standpoint of the scholar who wishes to use it as a tool for his work.

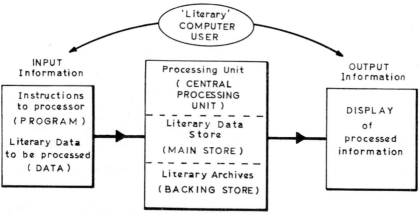

Figure 2.1 The 'Literary Computer' - block diagram of the processing of information.

The computer can be thought of as a collection of devices for storing and processing information. Modern computers code it in a digital form. Hence the title digital computer. It is coded in patterns of ones and zeros to match the pulses and non-pulses which can be provided by the electronic devices that make up the computing machine.

In our 'literary' computer, as in most contemporary machines regardless of their size, there are three main devices: (i) *input* and *output* (ii) the *memory* or *store* and (iii) the *central processing unit* (CPU). A fourth 'device' which must not, of course, be ignored is the literary user. Although the user might object to being called a device, in a modern interactive system the human user is an integral part of the system. In fact the user is encouraged to carry out a conversation with the machine. In the same way that the CPU or the memory reacts to a control signal, so too will the user react to the information that the machine presents. Unlike the electronic device the user's reaction can hardly be anticipated. The interactions between man (the user) and the various devices of our 'literary' machine are shown in Figure 2.2. Where the circle boundary in Figure 2.2 is a broken line it indicates that messages to control the various devices can be sent and received by the CPU's control unit.

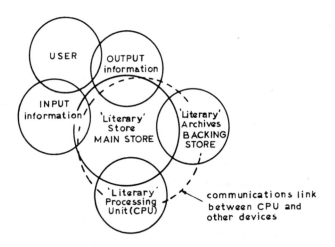

Figure 2.2 Interactions between the 'Literary Computer' User and the devices of the computer.

The input and output devices provide the paths whereby information is fed into the computer and is presented to the user by means of cards, tapes, teletype terminals and printers. Information is received from, and given to, the external world by these devices. Figure 2.3 shows some of the commonest devices now in use as well as mapping the relationships between the memory or store which retains information and the central processing unit (CPU) which processes it.

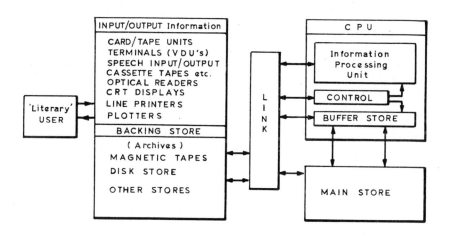

Figure 2.3 Literary Data Processing on a conventional computer system.

The CPU contains both a control unit and an arithmetic and logic unit, and it controls the operations of the whole computing system by issuing orders to other devices in the machine, and by responding to information it receives from them. The information stored in the memory can be either data to be processed or instructions to be obeyed.

Figure 2.4 illustrates the action of the computer when a typical set of instructions and data concerning a literary computation are presented to the input device. The set of instructions (called the program) tells the CPU which commands it must obey to process the literary text (called the data). Both program and data are input and stored in the memory, the CPU reads this information, interprets the instructions and executes them, carrying out operations on the data in accordance with the instructions and placing the results back into the memory storage. It arranges for the transfer of information between the various memory levels of storage and directs the information to and from the input/output devices.

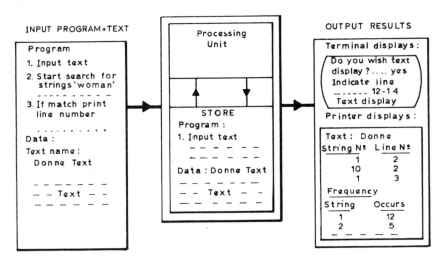

Figure 2.4 Action of processing a poem written by Donne :
The processor is instructed to search the text for occurrences of a given word or string of letters. Results displayed on a V.D.U. terminal and a lineprinter.

The computer's memory is now usually referred to as a store and the program and data are stored there before any computation takes place. There are various levels of storage determined by the need to access information at speed and the cost of the storage space itself. Figure 2.3 also shows the actual design of a typical computer. Most systems, whatever their size or capability, have an 'intermediate access' level of storage which interacts directly with the CPU so that information can be transferred at high speed. This is usually called a *buffer store,* and often consists of special storage registers. To provide this speed of transfer is costly and another level of storage called the *main store* is provided. This store is invariably larger and slower and much less expensive.

The lowest level of storage is that provided by the backing store which currently usually consists of *magnetic tapes* and *disks.* A *magnetic tape* is simply a plastic tape with information encoded on it, rather like the tapes used in the conventional cassette recorder. The *disk* has a surface in which information can be recorded in concentric tracks or grooves. The latter has been compared to the long-playing record or disk, and a unit consisting of many such disks is often called a *'Juke-Box'* store. Many other forms of backing store are now being developed. This provides much greater storage space, much of it in the form of exchangeable tapes or disks which can, if necessary, be stored away from the computer system. These storage devices form the *archive* for the literary computer user, which contains the library of his texts and programs. They remain stored in the archive until required. The results of processing can obviously be stored here as well as output for display to the user.

ACTION OF THE 'LITERARY' COMPUTER

Figure 2.5 also relates actual devices to the computer. These devices are realised in electronic equipment which is called *'hardware'*. It will be seen that the literary computer is a configuration of such devices with which the user can interact.

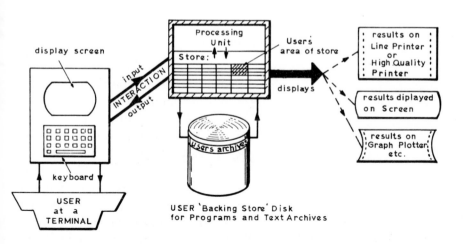

Figure 2.5 Information about a problem in literary studies presented to a computer using an interactive terminal.

Although these specific devices have been chosen for inclusion in the diagram because they are the result of current thinking and technological development, changes in technology can have a dramatic effect on design and construction of machines. For example, currently a speech input device is being perfected which allows the user to input his instructions to the computer using his own voice. Other new 'user-friendly' input/output devices, to allow pattern recognition, speech synthesis, and natural language processing, are being developed for the new fifth generation computers that are being developed by the international computer industry.

The action of the literary computer is one where information, usually a program of instructions and data, which could be a text to be processed, is submitted through one of the chosen input devices for storage, at one of the various storage device levels. This information is processed using the various levels of storage, by the central processing unit. It is this unit that reads, interprets and executes the stored information which represents instructions, obeying them in the sequence in which they are stored.

The sequence for a typical literary data processing problem would be:

Select the next instruction from stored program

Interpret or decode it

Select the data which is required

Obey the instruction

Continue until all instructions obeyed

The actual function of the instruction to be obeyed will depend on the machine designers and on the computer-user who has assembled the instructions for processing his data. The instructions made available by a particular machine are called the *'machine repertoire'*. They may include such commands as 'input data text', 'add a character', 'merge two patterns', 'shift a pattern of characters or letters to the left', 'clear out part of the store', 'obey the instructions held in a particular part of the store', 'add a number' and so on, covering as many operations as are feasible and desirable for the manipulation of the information patterns stored in the machine. A set of these instructions, the *program*, would be stored in some part of the store hierarchy and is called the *stored program*. The CPU has the task of obeying these stored instructions, automatically, in the sequence in which they are stored, and of performing the operations using the data information. Finally, it sends out the results of the processing to the output device and hence to the user, or perhaps the results would be retained within the system in the backing store which may be a disk file or on a magnetic tape.

SOLVING PROBLEMS ON THE COMPUTER

The literary user who wishes to solve his problems directly on the computing machine can code his instructions using the machine's repertoire. His set of instructions to the computer will tell the machine how the problem is to be solved. In effect, he is creating the sequence of patterns that are to be stored and ultimately interpreted by the CPU to produce the desired solutions. To do this, it is necessary to express the problem in a machine-orientated language. A degree of skill is needed to do this efficiently and effectively. Certainly it is time-consuming because the actual language of a particular computer has to be learnt before the problem-solving program can be written. If a user is not a computer expert he will usually avoid learning to program in a machine language. He will either use a previously prepared program or use one of the easily used problem-orientated programming languages that have been developed. Many of these languages have the style of natural languages and have been designed to allow easy access to the computer. Some have been specially designed for the literary user of computers.

Figure 2.5 shows how information about a problem in literary studies is presented to the computer in an interactive fashion, using a 'typewriter' type of computer terminal. Accessing a computer is a problem of communication and control involving man, the computer user, who wishes to solve his problem in literary studies, the machine and the information that he wishes to exchange and process. In chapter 3 we show how problems can be formulated and expressed in a programming language for solution by a computer.

SOME SIMPLE COMPUTER TASKS

Such is the ease with which instructions and data can be presented to the computer that the reader with no professional computer training can begin to 'program' almost immediately. Typical activities include:

Storing a Given Text

In most computer systems a text is stored as a named file in either the main memory or the archives. In some systems the text can be represented as the holes in paper tape or in cards. Data preparation equipment is available so that merely pressing the keys on a machine rather like a typewriter produces the punched tape or cards which are read into the computer by special tape or card readers.

Nowadays, however, by far the most popular method of storing a text is to type it directly into the computer using an ordinary 'typewriter' keyboard which is attached to a computer terminal. Many of these units have 'television-type' screens which display the text as it is being typed, and also allow easy alteration or editing. Many computer systems allow the screen to be partitioned and its various sections addressed separately.

Some of these systems allow the text to be typed directly to the archives—the backing store of the computer—and many 'key-to-disk' systems are available. The disk is the device used for the archive store.

Where a computer system is interactive and the user can have a dialogue or conversation with the machine the commands for storing a text usually appear on the terminal screen or typewriter.

The computer user types on the keyboard of his terminal which can display not only his typed commands but also the computer's replies. This popular type of terminal is called a visual display unit (VDU) and is available on most computer systems.

A typical dialogue between user and computer on this sort of system would be:

User (types on his keyboard a   :: Run File Creation System
request for the File Creation
System to be activated).
Computer (Replies displaying   :: File Creation System Accessed.
answers on the screen)   Name your File?

User types (to give a name to his file)	:: File Name = Donne
Computer (Replies, giving an invitation to type in the User's text line-by-line)	:: Type-in your File Donne
User (types in his text to be stored in a file, line by line—line numbers are generated)	:: 100 To make the doubt cleare that no womans true.

```
 200  Was it my fate to prove it full in you
 300  Thought I but one had breathd the purer ayre,
 400  And must she needs be falsee, because shes faire,
 500  It is your beauties marke, or of your youth,
 600  Or your perfection not to studie truth;
 700  Or thinke you heaven is deafe or hath no eyes
 800  Or those it has, winke at your perjuries
 900  Are vowes so cheape with women or the matter
1000  Whereof they are made, that they are writ in water;
1100  And blowne away with wind or doth their breath
1200  Both hot and cold at once, threat life and death
1300  Who could have thought so many accents sweet
1400  Tund to our words, so many signes should meet
1500  Blowne from our hearts, so many oathes and teares
1600  Sprinkled among all sweeter by our feares,
1700  And the devine impression of ...............................
      .........................................................
2000  Or must we read quite from what you speake.
      .........................................................
5100  And let his carrion corpse be a longer feast
5200  To take kings dogs, then and other beast.
5300  Now I have curst, let us our love revive.
5400  In me the flame was never more alive.
5500  I could begin againe to court and praise,
5600  And in that pleasure lengthen the short dayes
5700  Of my lifes lease; like painters that doe take
5800  Delight, not in made workes, but whilst they make.
      .........................................................
9000  .........................................................
9100  END OF TEXT DONNE
```

The text 'Donne' is now stored in the computer, with each text line numbered for easy reference in the file. It can now be retrieved from its store and displayed, with or without the reference line numbers, at the user's request.

A typical request for display would be:

User (Types at his terminal)   :: *Type-out File 'Donne'*

The whole of the text is then either displayed on the screen (a 'roll-over' operation called *scrolling* allows the text to be read, and any passage can be frozen for more detailed scrutiny), or it can be copied out on the computer's printer.

Editing a Given Text

Similarly an interactive dialogue can be held between the computer user and the computer so that the stored text can be edited. The line numbers given to the lines of text are now essential for reference. A typical editing session would be:

User (Types)	:: Run File Editing System
Computer(replies)	:: File Editing System Accessed. Name the file to be edited:
User (Types File name)	:: File Name = Donne
Computer	:: Start your editing
User	:: Delete Line 400
Computer	:: Line 400 deleted
User	:: Replace Line 2000
Computer	:: Line 2000 deleted and replaced by:
User (types in correct line)	:: 2000 Or must we read you quite from what you speake
Computer	:: Replacement complete
User	:: Insert additional lines 5900/6000
Computer	:: Type in insert from 5900
User	:: 5900: I could renew those times, when first I saw 6000: Love in your eyes, that gave my tongue the law. Insertion End.
Computer	:: Insertion complete New Lines 5900/6000 Original textlines renumbered from 6100

| *User* | :: End the Edit. |
| *Computer* | :: Edit Ended. New Version of Text-file Donne (1) stored. |

The new version Donne (1) edited according to the user's instructions is now stored in the computer. It is normal for the original file Donne to be retained in the archives until the computer is told it is no longer required. In this way, like the early proof copies from the printers' press, various states of a text can be retained.

Processing the Text

Instructions to process a text—called a program—can be presented to the computer on any acceptable media, for example, paper tape, cards, magnetic tape, cassette, floppy disks and so on. Where the computer is an interactive machine the instructions can be typed directly into the computer using a computer terminal in exactly the same way as a text was stored and edited.

In some systems the program of instructions is just another text to be stored and then processed by the central processing unit of the computer, which can decode and obey the stored instructions.

A program of instructions to process a stored text could be typed directly into the computer and executed, although many computer users still prefer to record their programs of instructions on various other media such as tape, cards and so on.

A typical program of instructions to process the text 'Donne' which we have already stored in the computer would take the form of a dialogue. The user would first type in a request to the computer to make available one of the computers' programming languages. The language chosen would be selected because it is convenient for issuing instructions for the processing of the text.

The dialogue would be:

User (Types in request)	:: Run 'Text Processing' Language System
Computer (Replies using display screen)	:: Test Processing System 'accessed'. Please type in your program name:
User (Types in his chosen program name)	:: Word Frequency

Computer (Indicates instruc-   :: Ready
tions can be typed in)
User (Types in his instructions ::
line by line)

                                  Shortened Form
1. Text file to be accessed :   1. FILES DONNE
    DONNE
2. Read a word from this text   2. READ WORD
    file
3. If the word read is the   3. IF WORD = 'END-OF-
    'end-of-text' marker then   TEXT' OBEY 8
    obey instruction number 8
    next.
4. If the word read is   4. IF WORD = 'WOMAN'
    'Woman' then obey instruc-   OBEY 6
    tion number 6 next
5. Otherwise obey instruction   5. OBEY 2
    number 2 next.
6. Record the occurrence of   6. WORD-COUNT
    the word 'woman' by   INCREASED 1.
    increasing the 'tally' of the
    number of matches.
7. Now read another word   7. OBEY 2
    from the text file by obey-
    ing instruction 2.
8. Copy out/display the   8. PRINT WORD-COUNT
    number of occurrences of
    'woman' noted by instruc-
    tion 6.
9. End of the Program of   9. END.
    instructions.
User (Types): Execute the   RUN
program

The computer will then execute the instructions of the program
sequentially, unless an instruction commands otherwise. Finally the
computer will display the count of the number of occurrences of the
word 'woman' in the text Donne. Some alterations to this program
would allow a full word frequency and concordance of the text to be
printed out by the computer.

The alternative form of the instructions – the shortened version –
given in the right hand margin is more typical of the current

programming language instructions available on most computers. Many of these programming languages allow this shortened version of our program to be written even more succinctly. The provisions of these languages and their usefulness to the user of the computer for literary studies will be considered in chapters 3 and 4.

SUMMARY

In this chapter we have introduced the computer and the concept of the 'literary' computing machine. We list some of the simple tasks that can be carried out using the computer, including storing, listing and processing a given text.

3. USING THE COMPUTER

The present state of the art in the field of computing is such that before a problem can be solved by a computer it must either be prepared in some considerable detail by the intending user or use is made of previously prepared sets of programs (often called software packages). In either case someone has to prepare the initial sets of instructions with great care if any confidence is to be placed in the solution.

There is, of course, a great deal of research taking place into the provision of semi-automatic or even fully automatic problem-solving interfaces and languages. Some of these developments have been directed at users working in the field of literary studies. The computer user and the computing machine can, for example, engage in a dialogue the aim of which is to help the user formulate his problem in such a fashion that it can be solved by the machine.

To appreciate the present means of preparation for the solution of a problem on a high speed digital computer we must realize that the following procedure has to be carried out, or has been carried out by those who have prepared the so-called software packages.

Setting Out The Problem

A clear statement of the problem to be solved must be made, usually by the user who works in the field in which the problem occurs. For a computer solution, however, the problem has to be formulated with the computer in mind. This means having some idea of the constraints of input and output, store, speed and efficiency, and so on, imposed when such a machine is used. In particular, the user should direct both his formulation and his proposed solution towards a target machine, that is the machine on which he expects to solve the problem. It is quite easy to be totally unrealistic about the power of present-day machines. Despite the technological advances of the age, and the capacity of some computers to perform millions of

22

operations in a second, it is not too difficult to present instructions to a computer which would not produce results until the next century.

Reduction of the Method to Simple Operations

The method chosen for the solution of the problem has to be reduced to basic operations that the machine can carry out, and the order of their execution planned precisely. Some of these questions will be arithmetical, others logical. Use is made of a *flow chart* or *block diagram* which is by far the most efficient method of representing this stage in a well-ordered graphical form.

One of the problems with using such charts is that, for their own internal purposes and for dealing with customers, each of the major computer manufacturers has over the years developed their own flow charting conventions. Various committees, user groups and 'standards' bodies have attempted to prepare standards, and have advocated them for general acceptance. For example, the American National Standards Institute (ANSI) standard consists of a set of graphical outlines or boxes which are called symbols. Some of these symbols are shown in Figure 3.1

Whatever standard is accepted, however, flow charts are fairly easily transferred from one format to another. The flow chart remains simply a set of boxes of various shapes containing the details of an operation and connected by lines called flow lines indicating the order in which these operations are to be carried out. The symbols shown in Figure 3.1 can, for example, be used to construct a flow chart (Figure 3.2) to represent the operations to be carried out by the computer to process the text 'Donne' referred to in chapter two. This chart shows how the word 'woman' is recognised in the text and a count made of its occurrence.

In the light of developing problem-orientated languages, many computer users believe that they can dispense with the flow chart. The authors would like to believe that languages will become so powerful and 'software packages' so easy to use that this will soon be true. But at present the 'state of the art' is such that there are few users who wish to prepare their own programs of instructions that can manage without the flow chart. There are certainly not many 'software' or 'programming' managers who would even contemplate its disappearance.

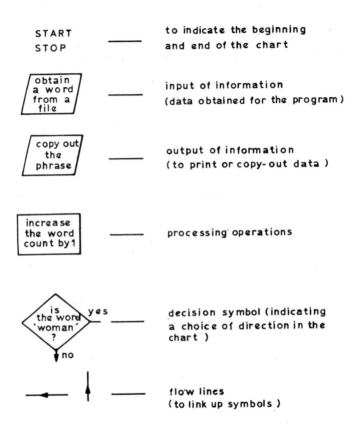

Figure 3.1 Some typical flow chart symbols

In setting out the operations to be carried out by the machine to solve a particular problem, a rule or method is being evolved for transferring a finite set of input data into a finite set of output data. In our example of Figure 3.2 the finite text 'Donne' is input, and the number of occurrences of 'woman' in the text is output as data. Such a rule is called an *algorithm,* and a flow chart is the means of expressing it symbolically or graphically. Algorithms must be well defined so that no attempt, for example, is made to compare an undefined value or to input more data than are defined. It, of course, also stops processing after a finite number of steps.

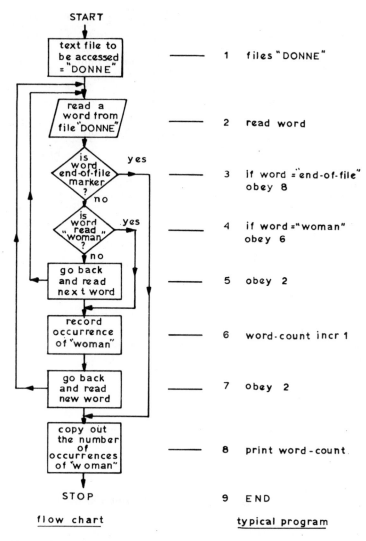

Figure 3.2 Flow chart listing operations with equivalent program of instructions.

Preparing the Computer Program

It is often frustrating for the 'literary' computer user who has prepared his flow chart in great detail for, say, a particular concordance task, to find that it cannot be directly fed into the machine and executed. Flow charting might well take on a new importance if future computer input devices will allow such direct input and storage of a chart for subsequent execution or translation into instructions that can be performed by the machine. The reverse of this process, that is producing flow charts from programs, has been achieved by many computer scientists. Indeed a systems flow chart language has been developed and implemented. Although instructing the computer directly from a flowchart may, therefore, become an option in the future it is now necessary to translate the flow chart into a corresponding set of instructions in a language the computer understands. The actual language of a particular machine, as supplied by the computer manufacturer, provides one means of expressing the instructions to solve a problem but most computers, as we shall see later in this chapter, accept many other languages.

The symbols or 'boxes' in the flowchart can be directly replaced by the instructions of these languages. In many languages there is often a one-to-one correspondence between the operation indicated in the flow chart 'box' and the language statements. This is illustrated in Figure 3.2 where each box has been replaced by the programming language statement which can be submitted directly to the computer. In fact the nine statements or instructions representing the flow chart would probably, as in the examples of chapter two, be typed using a teletype terminal (or visual display unit) into the store of the computer for immediate execution.

Because of the large number of *programming languages* available, the user has difficulty in choosing the best language for a particular problem. Many application fields have, for this and other reasons, developed their own specialist languages.

Transferring the Computer Program to the Machine

In addition to typing the program directly into the machine there are now a large number of different ways of communicating with the computer. Traditionally, the program statements or instructions were represented by the holes in paper tape or cards, scanned at high speed by the computer input device and sent to the computer's store as a set of patterns of 1's and 0's. It is, of course, possible to use methods involving speech input, graphics, optical readers and many other techniques as more and more input devices are produced.

Most computer systems usually provide a number of different methods of storing a program. Whichever method is chosen the program usually ends up in the computer's store in a named 'file', which can, if necessary, be corrected using editing facilities that are incorporated in the machine, before finally being executed by the computer system.

Testing the Program

The stored program is checked for errors and executed by the computer. Most computer systems offer a number of 'aids' to assist in checking first the language statements or instructions, and secondly, during the execution of the program, the 'run-time' errors which are monitored and information is presented to the user about possible errors.

Despite the research work that has been carried out for many years, it is still not easy to prove that a program or the solution obtained from it is correct. All that the user can do is to check the results of the computation most carefully against the known results for given test data. There are other techniques available to assess numerically the accuracy of the results obtained from a given set of data. It is unfortunately true that a very high proportion of computer users fail to realize that their initial set of instructions is transferred into another set by the machine before execution, and in addition that any computation that does occur is in the machine's numerical field, where approximations are made in accordance with the manufacturer's design criteria.

For the user from the field of literary studies, this process must seem rather a daunting one. Fortunately, in practice, using the computer has become a much easier task. A great deal has now been done to make the transference to the machine of a user's thoughts on how a problem should be solved, a much simpler process.

THE 'SOFTWARE PACKAGE' OR LIBRARY PROGRAM

Often the problems are such that the earlier stages we discussed above are passed through quickly and a program is written out immediately in one of the specialist or high level programming languages and then typed directly into the machine and tested.

Indeed the preparation is becoming easier as new generations of computers come into service, and new, easily used suites of programs become available. For many users in the literary field of study, the advent of more sophisticated programming languages has allowed them to miss out many of the main and time-consuming stages of preparation. For example, libraries of computer programs are now available world-wide. Many of the programs are housed permanently in the computer system to await data which is all they require for execution by the machine. A large number of programs exist in the literary field and a number of special-purpose *packages* or unified sets of programs have been developed over the years to meet the needs of the user. In addition to packages designed to perform such operations as word frequency, matching, concordances and so on, mathematical and statistical packages are also available on most computers. Many of these packages have been developed in specialized application areas and are designed for the user who is not a programmer.

Before actually writing new programs the user in the literary field is well advised to look to the literature and at the summaries of packages and programs that are now available. Most computer installations provide lists and the manuals which document their facilities. The more advanced systems have this information stored ready to be accessed by users who wish to enquire about any particular set of programs or packages available on the machine. Indeed many machines provide the information and summaries of the programs on VDU screens, and other machines are prepared to engage the user in a dialogue in an attempt to help him find the most suitable program or package for his particular application.

Several of the packages developed by Rudall [1] have been designed to provide the user with a man-machine interface that is easily mastered. The PROTEXT package [2] for example allows the following typical commands for searching. In this example an Anglo-Saxon text called WANDERER stored by lines in a file is searched for a given set of characters. All the user of the package need do to obtain the location of the words 'FORLET', 'CWITH', 'ENGEL' and 'GEDAEL' is to type the command to the package:

    SEARCH (BYLINE) TEXT WANDERER/
    PART 1 FOR 'FORLET',
    'CWITH', 'ENGEL' AND 'GEDAEL'.

The search routine in the package is activated and the named text searched and information about the given words output on the screen of a VDU or copied onto the line printer for a more permanent record.

Some of the packages available provide a dialogue entry. That is, they ask the user a series of questions about the problem to be solved, and based on the answers, select the appropriate routines to be carried out. In the author's CONTEXT package [3] for example, a typical dialogue between CONTEXT and the user who requires a limited concordance of five given words in a text called DEOR would be:

CONTEXT Asks	User Replies
⋆ Please name your text	'DEOR'
⋆ Has it a standard text format?	'YES'
⋆ Full concordance required?	'NO'
⋆ Concordance with exceptions?	'NO'
⋆ Selected concordance?	'YES'
⋆ How many selections required?	'5'
⋆ Name your selections?	CWAETH/WYRD/SAE/ARE/OT
⋆ Do you require left and right context?	'YES'
⋆ Concordance to be output to Line Printer?	'YES'

A concordance showing the location of selected words together with a left and right context (the words in a line appearing before and after the given word) would be printed on the line printer.

In later chapters several packages that have been particularly useful in the literary field will be discussed in some detail.

How much reliance can the user place on the results provided by a package? Of course, he can check his solutions, but in the main he has to rely on the package being rigorously tested by its originators. This is an area that is currently receiving a great deal of attention from researchers and it is hoped that a method of certification or 'package proving' will soon be forthcoming.

One of the most useful signs for the future of computing in the literary field is the very high degree of co-operation shown by users, who are prepared to share their programs or to submit them to one of the computer user organizations or associations for literary computing that maintain libraries of programs for general circulation. There is no doubt that communication with the computer is becoming more automated[4], and that slowly the stages of preparation presented here will be reduced, we hope, to stage 1, which only involved stating the initial problem to be solved.

SUMMARY

Using the computer to solve a problem requires that the problem be set out clearly and that the best method of solution be selected and reduced to simple operations before the computer program is prepared and transferred to the machine for final testing. Many problems can be solved by use of software packages or library programs, and some of these can be used in a 'question and answer' manner. The whole process of communicating with a computer is becoming much more automated.

NOTES

1. B H Rudall, 'A Cybernetic approach to the specification and interfacing of software modules', *Computers and Cybernetics*, London, 1981.
2. The first purpose built command language for processing texts: B H Rudall, 'A Command Language for Text Processing', *The Computer in Literary and Linguistic Research*, ed. R A Wisbey, Cambridge, 1971.
3. B H Rudall, 'Toward an Automatic Interface for Processing Systems', *Progress in Cybernetics and Systems research*, Vol. XI. Edited Trappl, Fidler, Horn, 1982.
4. Discussed by Karen A. Frenkel in 'Toward Automating the software-development cycle', *Communications of the ACM,* 28(6) (1985) 578–589, and also by B. H. Rudall, 'Towards Automatic Software Generation', *Robotica* 3(1) (1985), 31–34.

4. PROGRAMMING THE COMPUTER AND THE CHOICE OF LANGUAGES

THE DEVELOPMENT OF PROGRAMMING LANGUAGES

It is now appreciated that the computer performs its processing of information in a sequence of elementary operations which are available to it by virtue of its manufactured capabilities. This sequence of operations, so necessary for the computing machine to carry out its processing, is now universally called a *program* and the process of preparing such a sequence is called *programming*. All computer users dream that programming will, in time, become automatic, that is, that it should be possible to present instructions in a natural language to a computer that would automatically construct its own program of detailed instructions for execution. Indeed the computer system that can understand natural language is a highly desirable one and its development is closely tied up with research into the computer simulation of languages used by human beings, and to the research in artificial intelligence.

Historically, the preparation of instructions for execution by a computing machine has been based almost entirely on the hardware design of the computer. The architecture of the machine and the built-in repertoire of instructions implanted by the manufacturer were the prime considerations of (and constraints on) the potential user.

The first language developed for communication was consequently the language of the machine, invariably called *machine code*. The sequence of instructions produced in this code to make up a program was very closely connected to the machine's hardware. A very special knowledge of the machine's internal operations had to be acquired by a would-be programmer.

It soon became desirable to replace instructions which consisted of strings of digits by a symbolic representation, so that meaningful names, or at least mnemonics, could be used to represent the

31

machine's functions and store locations. This new method of representing instructions was called symbolic *assembly code*. It was still the language of a particular machine or machine range, but before it could be used it had to be converted back to the machine code by a special program called an *assembler*. There was still, however, a more or less one-to-one correspondence between an assembly code instruction and its machine code equivalent.

Many of the pioneering computer projects involving literary applications were programmed in the various assembly codes. Few users who were not professional computer programmers would attempt to master this method of programming.

Fortunately the advent of the new *autocodes*, which allowed the computer user a new, mathematical, type of language notation for the expression of his problems, simplified communication with computers. The language of the autocodes was problem-orientated and it placed the responsibility for the assignment of storage locations and for the conversion of the mathematical type of notation into machine functions on a specially prepared program called a *compiler* or *translator*. The idea was to develop a suitable language for communicating to the computer the problem to be solved. At the same time a compiler or translator was provided to translate the language statements into the computer's own machine code. Each autocode language consequently has its own compiler or translator which turned the computer into a more versatile and easily accessed machine.

Autocodes were soon called *programming languages* and IBM developed one particular language, FORTRAN, which was later to be almost universally adopted by computer users.

More sophisticated structured languages like ALGOL were designed and implemented in the late 1950s. Many were adopted as a programming language for literary users, and many of the initial programs in this field were based on them.

Developments in programming languages have now continued at a great pace. Instead of writing programs in machine code or the early assembly and autocode languages, the majority of computer users now program in a language sufficiently like English to be readable. These languages are international and some of the commonest in use in the field of literary computing are BASIC, FORTRAN, ALGOL 60, ALGOL 68, PL/1, PASCAL, SNOBOL. There are many others, some especially designed for applications in literary and linguistic studies.

Table 1

Some of the Higher Level Languages currently available

Numerical Scientific	Multipurpose	String Processing
*ALGOL 60	ADA	COMIT II
APL/360	*ALGOL 68	EOL-3
*BASIC	*EXTENDED ALGOL	*SNOBOL 4
EULER	HAL	TRAC
*FORTRAN	JOVIAL	VULCAN
JOSS	REL ENGLISH	ICON
MAD	OSCAR	
NELLIAC	PARSEC	
SPEAKEASY	*PASCAL	
	*PL/I	
	SIMULA 67	

List Processing	Artificial Intelligence
AMBIT/L	CONNIVER
L6	PLANNER
*LISP 1.5	QLISP
LPL	PROLOG
MLISP	
MLISP 2	Simulation
TREET	CSMP
	DYNAMO III
	MIMIC
	GPSS
	SIMSCRIPT I AND II

Computer Assisted Instruction	Editing and Publishing
COURSEWRITER III	CYPTER TEXT
FOIL	FILE COMPOSITION
LYRIC	PAGE
MENTOR	*SNAP
MRC	
*PILOT	Social Sciences and Statistics
1PLANIT	DATA-TEXT
TUTOR	ESP
	PROFILE
	TFL
	TROLL

* used frequently for the teaching and study of literature.

Table 1 lists some of the programming languages available. Some of these languages, although designed for an entirely different field of study, have been used successfully by users in the literary field. Even the universally used business language COBOL has been adopted for use in literary projects. The programming languages that are used frequently in the teaching and the study of literature have been marked with an asterisk.

TYPES OF PROGRAMMING LANGUAGES

Any set of instructions that can be executed by a computing machine may be termed a program, and the language in which the instructions are expressed is a programming language. Such languages can be classified according to a variety of criteria but a classification based on the *type of transformation* required to translate the language into a form that can be processed by the computer is particularly useful. Using this, languages could be divided into the three classes: machine code languages, low level languages and high level languages.

The machine code languages have a binary or numerical format and certain instructions, which usually correspond directly to these, are built into the machine's hardware. A typical sequence of machine code instructions is shown in Table 2. The first five digits of the instruction indicate to the computer's hardware the function it has to perform, whilst the remaining digits show where the data to be processed can be found in the computer's store. The low level languages are the 'assembly' type where mnemonic symbols and decimal numbers are used. One assembly language instruction would match each machine language instruction. The translation process being a simple one-to-one transformation by a program called the program assembler.

The high level languages like BASIC, PASCAL, ALGOL or FORTRAN that are used in the study of literature are problem-orientated and are much nearer to the natural languages used to formulate a problem. Programs written in the higher level languages are translated by a compiler or an interpreter which translates each high level instruction into a set of machine language instructions for execution by the computer.

This classification is illustrated in Table 2 where extracts from programs written in the three classes of language are shown. The main task of each program is to calculate the average length of lines of text stored in the computer. The number of words and lines in the text are counted and the final tally stored under the names Total-words and Numberlines. The Average is calculated by dividing one by the other. The various languages have particular rules for the way in which these names are used in programs. The language BASIC, for example, in many implementations, will only allow names to be represented by single letters or single letters followed by a digit 0 to 9. Hence Totalwords has to be represented by T and Numberlines by N. The actual arithmetical result will also depend on the particular language. Whilst some of the statements in the table will produce averages with a decimal part, others like SNAP and SNOBOL will produce whole numbers only.

Table 2

Examples of Programming Languages

(i) *Machine Code* *Binary*	(ii) *Low-Level Languages* *Assembly Language*	*Meaning of Operations*
11011 0 110111	L T	Load Total Words (T)
11001 0 100111	D N	Divide (T) Number of Lines (N)
10100 0 110011	S A	Store Average Line (A)

(iii) *High Level Languages*

BASIC	10 LET A = T/N
FORTRAN	AVLINE = TOTWRD/NOLINE
ALGOL 60 and PASCAL	Average Line: = TOTALWORDS/ NUMBERLINES;
PL/1	AVERAGELINE = Totalwords/ Numberlines;
SNAP	SET THE AVERAGELINE TO THE QUOTIENT OF TOTAL WORDS AND THE NUMBERLINES.
SNOBOL	AVLINE = TOTWORDS/NUMBERLINE.

Table 3 contains a program in BASIC which solves the problem outlined in the flow chart of Figure 3.2 in chapter 3. BASIC is an interactive language and the whole process of typing-in the program and of obtaining the results is indicated. The instructions are readable and match the flow chart symbols and explanations given in chapter 3. Whether BASIC is to be the language of your choice for work in literary studies is, however, another matter.

Table 3
Complete Program in Higher Level Language

Flow chart Instructions	Basic Program
1. FILES DONNE	10 FILES DONNE
2. READ WORD	20 INPUT 1 # W$
3. IF WORD='END OF FILE' OBEY 8	30 IF W$='*' THEN 80
4. IF WORD='WOMAN' OBEY 6	40 IF W$<>'WOMAN' THEN 20
5. OBEY 2	60 LET K=K+1
6. WORD-COUNT INCR 1	70 GOTO 20
7. OBEY 2	80 PRINT 'WORDCOUNT=',K
8. PRINT WORD-COUNT	90 END
9. END	

Note: Instruction 4,5,6 and 7 corresponding to the flow chart symbols have been rewritten in BASIC using the operation < > which means 'not equal to'.

1# merely refers to file number 1 which is called DONNE.

W$ stands for the WORD and K represents the WORD-COUNT and is zero at the start of the program.

* is the END-OF-FILE SYMBOL used to terminate the file.

The instructions have been renumbered in units of 10 to allow for the insertion of additional instructions, if required, at a later stage. Many versions of Basic allow the program to be written more succinctly.

A program to solve the problem of Table 3 using a higher level language of the PASCAL or ALGOL variety is shown in Table 4. Such a language is not restricted to individual statements or to the rigid format of the BASIC version that has been described. This program displays a structure, illustrated by the use of the symbols **BEGIN** and **END** which bracket statements together. Such a language is called a 'structured' higher level language.

Table 4
Pascal program to search the text DONNE for the occurrence of the word 'WOMAN'

PROGRAM   WORDFREQUENCY (INPUT,OUTPUT);

```
CONST                    WORDMAX=15;
TYPE                     WORDSIZE=1..WORDMAX;
                         WORDS=PACKED ARRAY
                         (WORDSIZE) OF CHAR;
VAR                      WORD,TESTWORD:WORDS;
                         WORDCOUNT:INTEGER;
PROCEDURE GETWORD;       external;
BEGIN                    WORDCOUNT:=0;
                         TESTWORD:='WOMAN        ';
                         GETWORD;
                         WHILE  WORD < > '*        ;
                         DO
                         BEGIN
                           IF WORD=TESTWORD THEN
                         WORDCOUNT:=WORDCOUNT+1;
                           GETWORD
                         END;
                         WRITE ('OCCURRENCE OF WORD
                             "WOMAN" =',WORDCOUNT)
END
```

Note: This is a PASCAL program which contains reference to a procedure (that is a set of instructions) called GETWORD which obtains a word from the text stored on the INPUT file named DONNE. This file is terminated with the symbol *. A word is read from the text DONNE and compared with TESTWORD which in this example stores WOMAN.

A count of its occurrence is stored in WORDCOUNT.

The operator < > means 'not equal to' and the Pascal symbols CONST, TYPE, VAR and PROCEDURE declare some initial information to the computer.

HIGHER LEVEL LANGUAGES

A higher level programming language will have the following main characteristics.

1. The language requires no knowledge of the machine code by the user and is significantly independent of a particular computer.
2. The language notation is fairly natural to its problem area and is not in some fixed tabular format.

Both of these properties make higher level languages particularly attractive to the users in the literary field. The first criterion means that programs can be transferred from machine to machine and indeed become machine independent. The last criterion is, of course, more subjective and has to be evaluated by the user in his own problem area. For example, of the higher level languages displayed in Table 2, the reader might judge SNAP to have the only natural affinity to the problem area.

From time to time the Association of Computing Machinery [1] reports on the currently existing higher level languages that have been developed or reported in the United States of America. A roster has been compiled of all such languages that have been implemented on at least one general-purpose computer, and should be in use in that country by someone other than the developer. A recent roster included 167 chosen languages. Every language is designed for a particular application area, although as we will see later in this chapter, the user may not agree with the developer as to its designated area.

In addition to the higher level languages mentioned, the user will also come across languages developed for communicating with operating and time-sharing systems, sometimes referred to as *command* or *job-control languages*. The large number of text-editing languages form a category of their own, some being extremely powerful in the hands of the researcher who requires to perform 'literary' data processing. So, too, are the many systems represented as data management, file management, query languages, information retrieval languages, question answering systems and management information systems. They are all examples of the development of the means of machine communication at a high level.

Although the designers of these languages have attempted to make them independent of any particular machine, users must be made aware that numerous additional facilities, such as structured programming or string or symbol manipulation added to languages like ALGOL and FORTRAN, useful though they may be, often make the language non-standard. Fortunately many of these languages are subject to standardisation by national and international standards organizations. For example, the status of the more important languages such as the US Department of Defence's computer language Ada, is the concern of the International Organization for Standardization (ISO).

Finally, it should be noted that languages must be implemented on existing computing systems, and this can prove expensive. It is also worth remembering before selecting a language for a particular project, that the US Association of Computing Machinery has reported the demise of as many as 27 different languages in a given year. This may well be considered to be progress, in that new and better languages are replacing the old and less efficient ones. Choosing a dying language may, however, be disastrous to a project.

CHOICE OF LANGUAGES FOR LITERARY PROJECTS

There is of course, no 'best buy' in the choice of a programming languages to fulfil the needs of any particular application in the study of literature. In many cases the chosen language may not be fully implemented on the target computers available. Even if a reasonable subset of the language is available, the translation process on the machine to be used may be inefficient. In many cases researchers have chosen low level languages because they produce efficient programs for execution on a given computer. The balance to be achieved in the choice of suitable language is often between ease of programming, in perhaps a fully structured higher level language, and the efficiency of the translation process in terms of translation time, together with the time taken to execute the final machine language version of the original program.

The error-reporting facilities made available in a language implementation are also an important consideration.

There are, however, new and improved languages being designed. Some of these will be considered in relation to the processing of non-numerical information.

The computer user with scientific or business applications, communicates with the computer in one of the many readily available higher level languages and the advantages of writing programs in such languages are now too familiar to reiterate. The user with applications in other fields is not so fortunate. Undoubtedly an area of the specialist, application-orientated language is with us, and the future heralds an increased use of computing machines by researchers in many fields. Meanwhile, however, the computer user engaged in literary studies has to make do, in the main, with languages which were not designed to meet his needs. Fortunately many symbol-manipulating languages have been designed and implemented and the researcher in literary studies who has expert programming assistance or the capacity to learn about programming, does have some aids. Even so, the programs will be written in terms alien to the field, and probably will not only be unnecessarily long and complicated but also slow in running.

Many of the higher level languages have been adopted for use in this field and FORTRAN in particular, has been well used by scholars. A good library of FORTRAN subroutines exists to perform many of the specialised operations required in literary research. Conversational programming and the use of microcomputers provides new horizons for the computer user.

The conversational programming language BASIC is one of the most widely used and well-known programming languages. Many 'literary' users have found that they can easily master it and use it conveniently to solve their problems, especially on microcomputers. It is an interactive, user-friendly language, which is simple in concept but its implementation invariably includes an editor, an error-finding system and easy-to-use operating commands. Many versions exist and many have been implemented to include some excellent facilities for the manipulation of files and strings. Recent versions allow structural programming in which simple statements are replaced by compound statements or even blocks of statements. Even so, BASIC is still regarded as a 'simple' language, essentially the equivalent of the languages designed for the first programmable 'hand-held' electronic calculators. The professional or the experienced user engaged in a major literary project usually scorns BASIC in favour of one of the more versatile languages. It has the advantage, however, of encouraging the novice to use the computer, even if, at a later date, he has to re-think his programming strategies.

The languages developed for numerical computation are already well known, but the literary user like many other specialist users, has still to be catered for properly. Finally, the literary user who has neither the expert programming assistance nor the capacity (or time) to learn programming languages has to resort to the use of the special software packages. Consequently there is a particularly strong incentive to researchers to produce more software which will meet the needs of users engaged in literary research.

LANGUAGES FOR NON-NUMERICAL COMPUTING

The computer is used to solve problems in many areas of literary studies. Where the problems to be solved are numerical in origin or are carried out on numerical data, they are more than adequately catered for using a computer which was traditionally developed with this in mind. Computing machines, however, are being used increasingly to process information which is non-numerical. The 'spin-off' from these developments, where symbols or patterns are processed instead of numbers, has provided programming languages and systems which can easily be utilised in the study of literature.

These languages have been developed to deal with the symbols or patterns that may well arise in programs to model the human brain, or its attempts at synthesizing and simulating intelligence, and in the areas of natural and artificial language research. The applications may be as diverse as storing and manipulating an image pattern transmitted from a robot, to processing the information contained in a business document.

In the main, the machines used for these purposes are still 'numerical' in function, and the patterns and symbols still stored in a numerical form. But because the information is really non-numerical, conventional numerical or scientific programming languages are seldom the ideal tool for manipulating these structures. This is precisely the case when such structures have to be processed in literary and linguistic applications.

Many of the original higher level languages like FORTRAN and ALGOL 60, have been extended in an attempt to cope with non-numerical problems. Newer languages like ALGOL 68, PL/1 and PASCAL have a multipurpose function, it is hoped providing facilities for the numerical and the non-numerical user. Remarkable as these languages have been in design and implementation, they cannot easily replace the purpose-built language which has been designed to enable the user to manipulate a stored data structure in a particular way. The data structure, such as a sentence written in a particular author's style, arises naturally in this form, and consequently it is desirable to preserve the structure in the computer's memory store.

The non-numeric languages developed over the years to deal with such structures include list processing and string processing languages. Other languages have been implemented to deal with specialized problems that occur in the particular areas of, for example, artificial intelligence, information retrieval, computer-aided instruction and so forth. The aim of these languages is that of providing the means for the description of the data structures and the operations which are to take place on them so that algorithms concerning the data can be expressed in a reasonably natural manner.

In the main application areas in literary studies it is possible to identify common structures and operations. One such structure, for example, is stored as a 'list' and operations concerning it can be carried out conveniently using techniques called 'list processing'. These 'list' structures appear for example, when a corpus is to be examined for certain properties. They occur frequently in problems concerning natural languages.

List-Processing Languages

These provide the computer user with computer memory cells that contain two items of information, the symbol or pattern to be stored and the address (or locations) of the cell containing the next symbol or pattern. In other words, the storage structure allows a linkage between the stored items, and the computer memory contains lists of symbols or patterns. By knowing the address (or location) of the first item in a list it is possible to gain access to the list. Any one list can be linked to another list and a whole structure developed. One of the obvious advantages of storing information in a 'list' is that the information need not be stored sequentially. A new item of information can be added quite easily, simply by putting a new item of information in any free cell in the memory and changing the link so that it refers to the cell containing the new information. A link is added in the cell containing the new item so that it refers back to the next item in the original list. Obviously the new lists, which may represent words, phrases, sentences and so on, can be inserted anywhere in the structure using this method. Figure 4.1 shows how the line by Tennyson, 'COME DOWN O MAID' is stored in the computer's memory and the new word 'FAIR' added to the structure.

A whole text, of course, can be considered as a list if its elements are to be linked into a structure that can be processed.

A whole range of operations are now made available in the list processing languages that have been developed. These operations allow symbols or patterns to be stored, compared, located and retrieved, combined, deleted and so on.

There is also a choice of list-processing languages available on most computers. The first work in developing these languages was in 1954 when Newell, Shaw and Simon developed IPL-V, an information-processing language. Subsequently, in 1960 McCarthy introduced LISP. The current version is LISP 1.5[2], which is a very sophisticated language, and is implemented on a large number of computers for both batch (the presentation of a complete program and data to the machine) and interactive working. MLISP is a language based on LISP but with an ALGOL style of syntax[3]. It is much easier to use in the preparation of problems. MLISP programs are finally translated back into LISP for their machine execution. Another version, MLISP2, which contains additional facilities, including 'pattern matching', is also available.

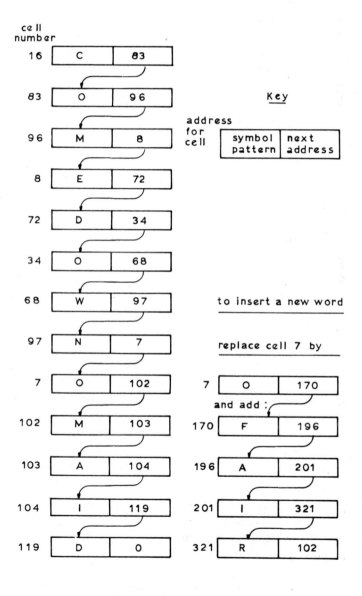

Figure 4.1 Storing a sentence as a list structure and inserting a new word.

String Processing Languages

These were developed in the 1960s and soon became a very popular and effective means of programming a computer for literary and linguistic research work. COMIT by Yngva[4], and SNOBOL[5] by Farber et al were among the first. All lists can be made into 'strings' and these languages may therefore be regarded as a subset of the list-processing languages. They are, however, designed for the restricted concept of the string and therefore more efficient within the area of string manipulation and storage. COMIT II is a version of the original string handling language COMIT, and SNOBOL 4 (of the original SNOBOL language) is available on a large number of computers. Other string-processing languages are available to the literary user, including TRAC(R) (Rockford Research Inc. USA) which is an interactive string manipulation language. One of the main purposes of using string manipulation languages is that such languages provide facilities for matching strings against a sequence of patterns. The pattern matching is performed by specifying the sequence of patterns to be matched. The patterns may be words, phrases, sentences or just a collection of letters or symbols. The matching of such patterns with the letters that make up a text could lead to the identification of a given section of the text, or could be part of a program to produce a word frequency list or a concordance.

Modern poets have indulged in pattern matching. An instance is Dylan Thomas. According to one of his childhood friends, Thomas constructed some of his poems in an unorthodox way. His aim was: 'To create pure word patterns, poems depending upon the sounds of the words and the effect made by the words in unusual juxtaposition . . . He carried with him a small notebook containing a medley of quite ordinary words . . . when he wanted to fill in a blank he read in his dictionary, as he called it, and tried one word after another . . .'[6].

This procedure is plainly reminiscent of sentence-generating programs. Part of such could be written in SNOBOL. Two of Thomas's lines could be defined as a string (denoted here by STR). The program would take the form:

STR = 'Because the pleasure-bird whistles after the hot wires,
Shall the blind . . . sing sweeter?'

```
STR = '. . .' = 'horse'/F (END)
SYSPOT = STR
END
```

Thomas's lines are searched for the blank '. . .' If it is not present no replacement is made. If the match succeeds, the word 'horse' replaces the blank '. . .' and the modified string which are Dylan Thomas's lines are printed: (the instruction SYSPOT = STR is a SNOBOL instruction to print out the string STR).

'Because the pleasure-bird whistles after the hot wires,
Shall the blind horse sing sweeter?'.

Professor Louis T. Milic[7] has used similar programs to reconstitute the poet's private dictionary and has attempted to create new Dylan Thomas poems. Such a process through the use of string manipulation languages lends itself to a high degree of computerization.

Nowadays sentence-generating programs are not new and many more programming languages have been developed for such non-numerical applications. The ICON programming language[8] is so designed, with a particular emphasis on string processing. It inherits the philosophy of SNOBOL 4 and that of another language SL5. ICON emphasizes expressive power in control structures. Expressions return a result consisting of a value and signal. The values are used in the normal computational manner, while the signals drive control structures.

COURSE WRITING LANGUAGES

The study of literature implies that the process of both learning and teaching will occur. Although this text is not specifically concerned with either as individual processes, the use of the computer in these applications is becoming increasingly important. The student of literature should note that there are many software packages and currently at least ten specialized non-numerical programming languages available to prepare courses of study for computer instruction. Probably the world's most advanced computer assisted learning system is PLATO, marketed by the US firm Control Data Incorporated. A language available for planning computer-assisted instruction courses to run under this system is called TUTOR[9].

Other languages like PILOT[10], MENTOR[11], and PLANIT[12] often referred to as 'course writing' or 'Authorship' languages[13] are available to prepare programs for most computing systems so that they can be used for assisting in the study of literature and linguistics.

LANGUAGE DEVELOPMENTS

With the advent of new technologies, the emphasis on new computer architectures, the new 'user-friendly' interfaces, and the introduction of artificial intelligence, there is a move away from the traditional higher level languages towards languages such as PRO-LOG and LISP. These languages have enhanced symbol manipulating and logic programming facilities. They provide the means of developing problem-solving and interface making software, and are closely involved in the development of the new generations of 'knowledge processing' supercomputers.

PROLOG (PROgramming in LOGic) has been adopted as the kernel language for the fifth-generation computers. It makes use of a form of basic mathematical logic and allows problems to be solved by means of a series of logical inferences.

The full implications of these developments to the literary user have yet to be evaluated, but there is no doubt that the means of specifying information processing tasks will undergo a change. Indeed a specification of a problem in PROLOG can, using formal techniques, be proved for correctness.

There are many derivatives of the language which include Micro Prolog for microcomputers, MPROLOG and DURAL. At the same time as computer architectures change, new languages like VALID, which has been developed for programming computers called 'data flow' machines, are being developed. VALID-S, for example, has been specifically designed for symbolic manipulation.

This is by no means an exhaustive list of the languages which are becoming available to the literary user. Indeed the software and hardware of the new generations of computers will allow the user to communicate with machines in a variety of different ways, and probably through an intelligent interface, which could well involve the use of natural language.

SUMMARY
Programming the computer to solve a problem involves a choice from a wide range of programming languages, which have been classified into machine-code, low-level and high level languages. A wide selection of languages is available for literary projects and many high level languages have been adopted for such applications. For many of the problems involving symbols or patterns, special languages for non-numerical computing, including list-processing and string-processing languages, have been developed, and for learning and teaching application in literary studies, many specialist course writing languages and software packages are available. For the new generations of computers, new languages and 'intelligent' user-machine interfaces are being designed.[14]

NOTES

1. See 'Roster of Programming Languages', published at frequent intervals, by The Association of Computing Machinery.
2. Original manual by J. McCarthy, *LISP 1.5 Programmers Manual*, Cambridge, Mass, 1962.
3. D C Smith, *MLISP* Computer Science Dept. Rep. No. CS-179 and *MLISP2* Rep. No. CS-73-350, both by Stanford University, 1970 and 1973 respectively.
4. Original description by V H Yngve, *COMIT Programmer's Reference Manual and Introduction to COMIT programming*, Cambridge, Mass, 1961.
5. Introduced by D J Farber, R E Griswold and I P Polonsky, 'SNOBOL: a string manipulation language' *J. ACM* II, No. 1, 1964.
6. J H Martin, Correspondence in *Times Literary Supplement*, 19 March, 1964, p.235.
7. Discussed by Louis T Milic 'The possible usefulness of poetry generation', *The Computer in Literary and Linguistic Research*. Edited R A Wisbey, Cambridge, 1971. See also M D Fosberg Harris, 'Dylan Thomas, the Craftsman: Computer Analysis of a Poem', *ALLC Bulletin*, 7, 1979.
8. Described by R E Griswold and D R Hanson, *Reference Manual for the ICON Programming Language*, Computer Science Dept. Tech. Rep. TR79-1. University of Arizona, 1979.
9. B A Sherwood, *The TUTOR Language* Computer-based Education Research Lab. Rep., University of Illinois, 1974.
10. Introduced by S Rubin, 'A simple instructional language', *Computer Decisions*, 5, 1973, and now available on a large variety of computers.
11. Outlined in: *Computer Systems for Teaching Complex Concepts* Final rep. No. 1742, Bolt, Beranek and Newman, Cambridge, Mass., 1969.
12. Described by S L Feingold, 'PLANIT—a flexible language designed for computer human interaction'. *Proc. AFIPSFJCC*, 31, 1967.
13. 'Coursewriting' languages are discussed by B H Rudall and J A Secker in 'Developing Automated Teaching systems', *Cybernetics and Systems Research*, Edited R Trappl, Amsterdam, 1982, and their generation by B. H. Rudall, in *Cybernetics and Systems*, edited R Trappl, 1986.
14. Two useful guides to popular languages particularly written for Humanities programmers are Peter Adman, *Fortran 77 for non-scientists*, Bromley, 1984, and Susan Hockey, *SNOBOL Programming for the Humanities*, Oxford, 1985.

5. USING THE MICROCOMPUTER

NEW DEVELOPMENTS IN MICROELECTRONICS

The new technological developments in microelectronics have profoundly affected our approach to computing machines. Quite suddenly, the literary computer user has access to smaller, faster, more efficient, more reliable, more versatile and less costly computers.

The traditional computer, the 'Von Neumann' computer, will clearly remain in operation for many years. But many of the early hardware constraints on the design of computers have been removed. The availability of the new microelectronic technology will enable us to rethink the way in which the user communicates with the single central processor. Indeed, the user could find that he has at his disposal not a single processor but 100,000 or more such processors, and the whole concept of his relationship with such a machine will have to undergo radical change.

This new technology has achieved a complete revolution in the production of hardware. For example, the capacity of a large central processor of a computer system can now be packed into a chip of semiconductor material measuring less than 1 cm^2 in area. The store used to retain the texts and programs to be processed can be inscribed on the same chip as the processor. New techniques are making it possible for more and more to be put on a single chip. The computer has become a lattice of crystals, with the scientist striving to put more processors and larger memory units onto them. At the same time he is experimenting with new and more versatile materials from which chips can be manufactured.

Microelectronics has enabled us to design and manufacture computer systems in a revolutionary way. What is most remarkable is that not only do we have the capacity to design more powerful and more accessible computer systems, but that we can also achieve this more cheaply than with the traditional electronics of the earlier machines. The chips of semiconductor material, currently silicon, are becoming astonishingly cheap.

The personal computer has arrived to provide the user with his own computational facilities. Such a system comes complete with computer programs which enable it to be run by non-specialists. At the same time, more expensive computer systems are being developed, using microelectronics, to replace existing minicomputer machines. The larger machines, the mainframe computers, are also affected. New designs based on microelectronics are being perfected, but in the meantime designers are replacing existing computer components with new microelectronic devices. They are doing this not only because they are much cheaper, but also because they are more reliable, use less power need less space, and require less air conditioning[1].

THE MICROCOMPUTER

Whatever the future, the microcomputers currently produced are based on the designs of the traditional computers described in earlier chapters. But some explanation of a few of the new terms used in association with these machines is necessary. Indeed, it is not possible to purchase a microcomputer with confidence without a little knowledge of the range of options on offer.

Firstly, the central processing unit (discussed, in the context of conventional computers, in chapter two), in the case of the microcomputer, is reduced in scale to fit exactly onto a chip or part of a chip. Such a chip or part of a chip is called a microprocessor. Unfortunately, the term 'microprocessor' has also been used since the advent of microelectronics as a popular name for the whole combination of devices which make up a computer. When a microprocessor is assembled with a memory and various interfaces for the input and output of information the assembly is called a microcomputer. Such a microcomputer could typically be assembled in an area less than this page. Complete computer systems are now being built on a single chip.

The main component of the microcomputer is the microprocessor, and throughout the world different designs of microprocessor chips are being manufactured. Associated with the microprocessor is the store or memory system of the microcomputer. Such a system must provide for the rapid storage and retrieval of the digitized information.

When dealing with a microcomputer the user should appreciate that these memories can be either 'moving-surface' devices or purely electronic. Moving-surface memories allow information to be stored in particular regions of some thin magnetic material. Surfaces may be flexible, such as plastic tape or disk, or else rigid, as in other kinds of disk. One of the most popular methods of storing information for microcomputers is the cassette tape recorder where the cassette or cartridge can store over a million individual items. It allows access to this information in times ranging from seconds to as long as two minutes but is now obsolescent. Flexible disk storage is much more widely used and may allow for the storage of many millions of items of information. Other storage devices, based on laser technology, have been developed and will transform storage capacities and access times, allowing even greater versatility.

The microelectronic memories allow thousands of items of information to be stored electronically on a memory chip. One such memory is called random-access memory (RAM). RAM chips are currently being produced to contain from 16K (that is, 16,000) items of information, and chips which can contain 256K (256,000) and more are often standard equipment. Indeed, we are told that the multi-billion-item chip is a distinct possibility at some future time. Many microprocessors require the program of instructions or data to be processed to be stored permanently. Into this category come the sets of instructions used by most pocket calculators. This form of storage is usually in a random-access memory which only allows the information to be read when required. Such devices are called read-only memories (ROM) and they contain permanently stored information placed on the chip when it is made. Thus, if a microcomputer is to be used for processing a text, the text (or perhaps only part of it, if a small store is being used) is conveniently stored in the RAM, while the software programs or package of instructions is stored in the ROM. Many of the complete suites of programs or packages produced by the manufacturer are housed permanently in the ROM. In some microcomputers, however, this information, more conveniently, is stored on a disk (either of the rigid or the flexible type) and brought into the RAM as it is required. If, for example, a user requires the microcomputer for BASIC programming, then it is likely that the ROM containing instructions to allow the computer to understand BASIC (that is the BASIC translator) must be obtained and linked to the computer system. Many different sets of packages and programs for use in literary studies can be housed in this way.

The microcomputer, then, consists of a microprocessor (acting as one central processing unit), memory devices, and various interfaces for the input/output of information. Figure 5.1 shows in block-diagram form how a microprocessor may be associated with the other devices to form a 'basic' computer system. This diagram follows the design of the 'classic' computer shown in Figure 2.1 of chapter two. Figure 5.2 shows how the user interacts with the microcomputer system. Such, however, is the variety of microcomputers reaching the market that no standard architecture exists, though the main components we have described can still be identified in most systems. With the present rate of innovation it will be increasingly difficult to pick out common features. Currently the types of computer chips described can be mounted on a board called a 'printed circuit board'. This is a sheet of plastic in which are

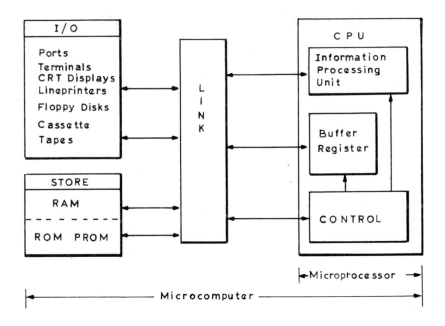

Figure 5.1 Typical microcomputer used in literary studies.

I/O - Input/Output devices   ROM - Read Only Memory
RAM - Random Access Memory   PROM - Programmable
CPU - Central Processing Unit   Read Only Memory

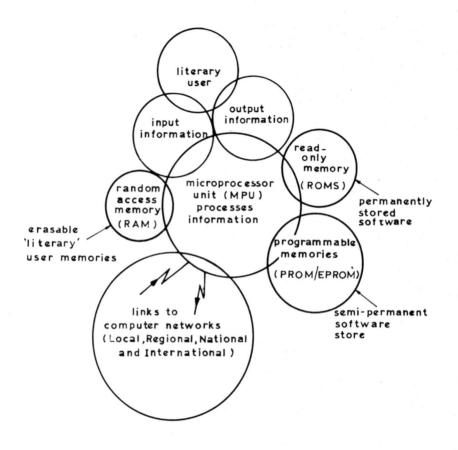

Fig5.2. Interactions between a 'literary' user and a
microcomputer system which can communicate
with the national and international computer
network.

embedded patterns of electrical wires which enable the chips to be
linked together. Figure 5.3 illustrates a board which contains the
microprocessor unit (MPU) for processing and control, while two
different kinds of memory chips provide the main store. This forms
a complete 'plug-in' computer processing unit.

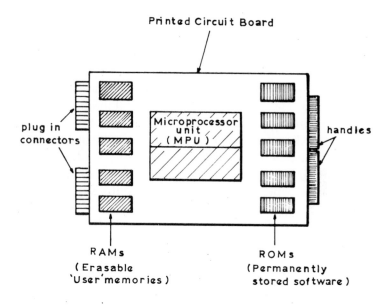

Figure 5.3 Diagram of a printed circuit board
containing the microprocessor unit
and memories . Together they form
a complete plug-in computer
processing unit .

USING THE MICROCOMPUTER

The user with applications in literary studies will use the microcomputer in much the same way that the larger, mainframe machines are used. If the user accepts what is sometimes called the 'black-box technology' approach, regarding the machine as a sealed unit, the internal workings of which need not be considered, then most of the work will be presented to the computer via a terminal of the type, already familiar from the mainframe machines, and prepared packages of programs will be used to resolve the problems presented for solution. Using the large, mainframe machines often presupposes the assistance of a professional staff to operate the machine, to deal with the storage and archiving of texts and the printing of results, and for the general maintenance of both the machine's hardware and software. This sort of 'back up' is not always available when a microcomputer is used, although the machine is likely to have its hardware maintained professionally on a maintenance contract. The software is usually 'bought in' by the computer user, who, in the case of the personal computer, is often the owner of the machine. Much of this software is modular and easily packaged and maintained, but some knowledge and experience in software handling is needed.

There are perhaps three approaches that the student of literary studies who needs a microcomputer can take to the machine. Firstly, there is the 'black box' approach, where the machine is so programmed that the user needs to know little about either the software or the hardware being used. Such 'user-friendly' systems are available, but a great deal of development is needed before they can be considered 'foolproof'. In the second approach, the user is prepared to read the manuals which describe the working of the microcomputer and indicate how the software package can be mounted, how programming languages can be used, and how the filing system and editors can be accessed. In addition, the user must be prepared to use the devices of the machine in a practical way, mount the disks, reload the printer with paper, and so on. In the third approach, the user is prepared to develop almost a professional expertise on the machine, becoming knowledgeable about its hardware, proficient in its programming languages and operating systems, prepared to test and run the software provided and generally operate the system. It is surprising how many students of literature have become students of personal computing as well. Whether it is necessary to cultivate this expertise is debatable. It is certainly the aim of computer designers to produce computers that can be used successfully by the 'non-professional' user.[2]

Some of the differences between using a microcomputer and the larger mainframe for processing applications in the literary field should be noted:

Input and Output of Information

Although the input and output devices which are available for the mainframe computer can also be used with the microcomputer, it is more likely that information will be input using the teletype keyboard and will be output using the screen of a visual display unit or on a printing device called a 'hard copier'. Magnetic tapes and flexible plastic disks, called 'floppy disks', have proved to be the most popular media for storing information, and in contemporary microcomputer systems disk drives, which are now cheap and very reliable, are essential input/output devices, cassette tape recorders are rapidly diminishing in popularity and by 1987 are likely to be very much a thing of the past. Texts for processing and programs of instructions are most frequently stored in these media for input, and the results of a computation can be copied out onto a magnetic tape or disk.

Storage of Information

One of the main characteristics of the mainframe computer is its capacity to store large quantities of data and to facilitate its retrieval. Although the first microcomputers did not have such capacity, as we have explained, many contemporary microprocessors have very large RAM memories, and this is currently a major area of development. However, though these technological advances are well underway, the literary user of microcomputer will probably need to use auxiliary storage devices. These will sometimes, for reasons of cost and compatibility, still include cassettes of magnetic tape as well as floppy disks. The cassette tape recorder is fairly cheap, but this method of storage does have grave limitations. Programs and texts are stored sequentially and the computer must search through the tape until it finds a specified item. When a program is stored on the tape it is usual to ask the computer to verify its recording. Disk drive units offer several major advantages. When a particular item is requested, the unit checks an index which it keeps on the first track of the disk. When it finds where the information is located, it goes directly to that area without searching through any other information. This direct method of accessing information is called random access. Another advantage of using disk units for storing information is that a floppy disk can hold far more material than a tape cassette.

The Microcomputer's Capability

Microcomputers are extremely versatile and can be modified to improve their performance and capabilities much more easily than mainframe machines. The student of literature can, for example, purchase a software program as required and can read it into the machine whenever it is needed. Many such programs can be purchased fairly cheaply and include such functions as 'word processing', 'graphics', 'statistics', and so on. Many computer organizations sell ROM cartridges containing particular programs to fit into the machine and act as a permanently accessible store of software. Most programming languages are available because the translators or compilers can now be obtained on ROM chips. Structured programming languages are commonly available, and versions of languages, originally designed for the larger mainframe computer, can be obtained. The software libraries are likely to grow at a great pace, since perhaps a million or more personal computer users may be involved in developing programs and packages for their machines. For example, several sophisticated concordance packages are, or soon will be, available for microcomputers.

There are, however, a number of caveats. Though the power and versatility of microcomputers have increased and are increasing rapidly, even the most advanced is less suitable than a mainframe for some literary applications, such as concordance work on large texts. Secondly, a much wider range of software is available for some machines (generally the more popular within each price range) than for others. This may be pertinent when a system is being selected. Finally, some systems can more readily be enhanced than others, and purchasers should consider whether the system they are buying can grow to meet probable future applications.

Word Processing

Although most mainframe computers have a system for word processing, microcomputers have largely taken over this role. Most microcomputer companies offer software programs, such as Wordstar or Easywriter, which give the user much greater control over what appears on the screen.

In essence, these systems allow users with no experience of computers to use the microcomputer screen as a writing pad where words are inserted, sentences arranged, texts edited and manipulated, to create new copies on the screen, before finally being printed by the machine. Drafts can, with facility, be updated to incorporate revisions and reformatted to fit almost any size of page. Copy can be produced with right hand justification, and if the printing is done on a high quality printer, then texts may be produced sufficiently well to be photographed for the press. It has been argued that word processing is particularly useful for literary editing [3]. Texts can be keyed in and corrected with facility and collation programs invoked to identify variant readings. But the chief advantage comes in preparing the edited version of the copy text. Alterations can be introduced very easily. Suppose that the edition being produced is to be normalised, and the editor has forgotten which alternative has been adopted earlier in the text. Word processing programs have the facility to search former portions of the text and resolve the problem. Again, if a recurrent normalisation is to be made throughout the text, word processing programs permit all the changes to be made with just one command.

The Dual Role of the Microcomputer

The microcomputer can act as a link to other computing machines. In particular, a microcomputer can be adapted to be a terminal to a much larger computer (Figure 5.2). The literary computer user can work on a microcomputer at home or in his office, but, if resources, such as greater processing power, an extensive data base or a special package is needed, a link can be made, and communication established, with a larger machine, perhaps at a very remote site.

SUMMARY

New developments in microelectronics have produced the microcomputer, which provides the literary user with a much more compact, reliable and less costly machine. Using the microcomputer is largely similar to accessing the larger mainframe computers, but there are differences concerning the means of input, output and the storage of information. The microcomputer is extremely versatile and has the capacity of being enhanced to provide an improved performance and new facilities. Word-processing is rapidly being taken over by the microcomputer and many manufacturers provide specialised word processing software. The microcomputer can also play a dual role, acting as a convenient link to other machines, and can thus usefully be used as a terminal to larger computers.

NOTES

1. The effects of the microelectronic revolution on the digital computer is discussed by B H Rudall, ('Microelectronics and the Cybernetician', *Computers and Cybernetics*, London, 1981.
2. For assistance in the choice of microcomputers, see 'The Benchmark Reports', published by the Association of Computer Users (ACU), Boulder, Colorado, and the analysis in 'The relationship between benchmark tests and microcomputers price', by Sircar and Dave, *Communications of the ACM*, 29(3), 212–217, 1986.
3. George Guffey, 'Microcomputer Program and Data Bank Projected for Editors of 17th and 18th Century Texts', *The Clark Newsletter, 2, 1982, 1–3.*

6. THE PROCESSING OF NATURAL LANGUAGE

CONCORDANCES

The Functions of Concordances

A concordance is a research implement, not an end in itself. There is no inherent value in possessing, say, all the words in *Hamlet*, rearranged in alphabetical order and surrounded by pieces of context, when already we possess Shakespeare's version—unless, of course, the scrambled and sorted text can be used to facilitate or refine investigation into the original. As we shall see, a concordance may assume any of a variety of forms. Defining its purpose, in terms of who will use it and to what end, is the first step in concordance production.

A concordance effects access to the original text in several ways. Most simply, it can function as an aid to the memory. A reference, a quotation, an allusion, perhaps only half remembered, can be checked in the concordance and thus located. Similarly, it can be used as an index, though better indexes, as we shall see later, can be produced using refinements of concordance-generation. Concordances facilitate several established approaches of linguistic stylistics to the study of both lexis and syntax. Such listing of the words of a text can make it much quicker to pick out and classify the compound words, and isolate possible neologisms or archaisms or slang words, which can then be looked up in appropriate dictionaries, to classify a text as 'using a lot of classical loan words' or 'drawing heavily on the native vocabulary of English' or as belonging to any of the similar categories to which critics have customarily assigned texts. A concordance has clear advantages over simple word-lists since it permits the user to distinguish homographs, which, as we have argued earlier, is a necessary stage in parsing which the machine cannot with facility effect[1]. If one's purpose is to discuss the syntactical distribution of a text (how many pronouns? how many adverbs? and

59

so on), a concordance greatly accelerates the process. Accidence can be investigated. It is feasible to have items listed in alphabetical order based on the end, rather than the beginning of words. Concordances, too, may be used in thematic analysis, in the identification of clusters of associated words which distinguish the linguistic universe of the text. A critic may wish to identify groups of words associated with light or money or sex or whatever. Similarly, one text or one part of a text may be compared with others. Does, for example, Macbeth inhabit a different linguistic universe from the other characters in the play? A first step might be to generate concordances for Macbeth and for the rest of the characters (no difficult matter, as we shall see) and compare them. Does the sort of words Milton uses in his pastoral poetry distinguish it from his epic poetry? Again, a concordance which observes this distinction will help.

A concordance may be generated by a solitary research worker, intended for his eyes alone, and designed as an instrument to solve a precisely defined problem in his project. Alternatively, the concordance may be a work of public benefaction, an opus produced by its maker not specifically for his own use, but dedicated and donated to his fellow workers in a chosen field, a research tool to be used by many and for a multiplicity of purposes. The former may reflect the idiosyncracies of its maker. Nor need its presentation matter. It may be listed like ordinary computer output on lined or perforated paper. It may even be kept as a computer file on disk or magnetic tape, to be consulted purely by interrogating the machine. In contrast, the public concordance, to be as useful as possible, must be produced to a saleable or at least publishable quality, fit for book publication or perhaps microfiche distribution. The private concordance, since it is probably produced in only two or three copies, may well be full and exhaustive. All words may be listed, surrounded by very generous contexts. The public concordance is shaped by the exigencies of publishing economics, and its maker will have to consider ways of controlling its bulk.

Programs and Packages

Concordance generation is familiar ground for computer science, and several programs have been developed which can be adapted to the requirements of particular projects. In this and some of the sections which follow we shall frequently refer to the two best known, COCOA and OCP. Both are flexible packages of programs which have achieved a considerable distribution. The latter, the Oxford Concordance Program, has been available since the early 1980s. It is very widely implemented in the United Kingdom, North America, and elsewhere. It was designed in part to supersede COCOA, which was developed in the late 1960s at the Atlas Computer Laboratory, Didcot—its name is an acronym from COunt and COncordance on Atlas. A comparison of the programs is particularly instructive, in that it illuminates the greater user-friendliness of recent software and points the direction for future developments. No doubt the next generation of programs will permit the user to define fully his problems and requirements through some very easy interactive procedures. Both COCOA and OCP are well documented [2], and their manuals should, of course, be carefully consulted before projects are designed and material prepared.

COCOA has both the advantages and disadvantages of its considerable antiquity. It cannot produce lower-case letters, and is far less flexible in its specification than OCP. The commands to be mastered are opaque and remote from natural language. To produce even a very straightforward concordance, a full, centrally aligned concordance with left overflow suppressed, the user must construct the following commands, each one punched on a separate punch-card (or the operation can be performed at a terminal):

```
(card  1)   72
(card  2)   0<,.;:?!>
(card  3)   1<'->
(card  4)   2*
(card  5)   3<ABCDEFGHIJKLMNOPQRSTUVWXYZ>
(card  6)   4*
(card  7)   5*
(card  8)   6*
(card  9)   7*
(card 10)   W<1 24>
(card 11)   P<T4>(<TEXT>)<ZZZZ>1000000
(card 12)   C<E5P3L3>100#100 C(A Z/1 99999)
(card 13)   +
```

This may seem like an arbitrary invocation to a strange god, and indeed, it does require quite a bit of effort to learn how COCOA commands work. Consider, for example, card 11. This is the instruction to process (P) a section of the file (in this case, in fact, the whole file) that has been marked with the reference code <T TEXT>. Since the whole text is to be processed right to the end, the card tells the computer to proceed until it encounters the reference code <T ZZZZ>. This is a dummy reference. No <T ZZZZ> is embedded in the text, and so the machine will process the whole file. The term '1000000' specifies the number of lines to be processed. Since we require a concordance to the whole text and since, perhaps, we know its length only approximately, we fill in a larger number than our estimate. We discuss this example, not to equip the reader to use COCOA—of course, for that the documentation must be read thoroughly—but to suggest the sort of problems to be met. These commands are rather inelegant. It is clumsy to have to tell the machine to search for dummy elements. The commands are remote from natural language and difficult to remember, and their syntax, with the use of various kinds of brackets and the precise spacing, must be meticulously observed. Errors are easy to make. However, the problems are not insuperable, and there is no reason why the critic or scholar should not master all the resources of the COCOA package with very little assistance from the computer scientist.

A set of OCP commands to perform a concordance-generating, listing or indexing task may consist of four major sections:

> *INPUT (This defines the format of the input text and specifies which parts of it are to be processed).

> *WORDS (Especially useful in processing foreign languages, in that it permits the specification of groups of characters which are to be treated as single letters. This section may also contain some statement of the longest word to be searched for, which can effect economies of computer time).

> *ACTION (specifies the sort of processing procedures which are required and the format for references).

> *FORMAT (Determines layout of output, lines per page, and so on, defines the configuration and layout of contexts, references and headwords required for each word in a concordance).

The OCP system, however, offers sensible and appropriate default options, so it is not necessary to introduce a command for each of these sections. To produce a very simple concordance, for example, we need only command:

```
*ACTION
DO CONCORDANCE
*GO
```

Text Selection and Preparation

The researcher who is making a concordance for his own use enjoys certain distinct privileges. He may select any version of the text that suits and seems valid for his purpose, and he may alter it at will. English texts written before the eighteenth century often exhibit considerable orthographical variation. The same word may occur in several forms, and, if the raw text is simply processed by a concordance program, should none of the measures discussed below be taken, orthographically distinct instances of the same word will be listed separately. Such a concordance, say, to *Paradise Lost* would separate the examples of 'Glorie' from those 'Glory' by the entries for 'Gloried', Glories', 'Glorified', 'Glorify', 'Glorious', and 'Gloriously'. This may seem an unacceptable nuisance. The private concordance-maker may silently modernise or at least normalise his text as he transfers it into a machine-readable form, even though it is of less authority than the edition he would use for other purposes of academic research.

The public concordance-maker has two basic alternatives open to him. He can adopt as his base-text a widely current and academically respectable edition[3]. This is a sound policy with much to recommend it. The concordance user will often be consulting it to find a reference in the master text, and, plainly, it is helpful if the references he finds are to an edition available to him. The second alternative, basing the concordance on early printed versions or on manuscript, requires rather more preparation time. Of course, for some writers there is no authoritative modern edition in general currency. However, there is a sort of season to literary studies: generally, the editor precedes the concordance-maker. The best reason for going back to originals is to allow the concordance to be used for purposes which would not otherwise be possible, such as an appraisal of old spelling (which sometimes offers ambiguities lost in modernisation) or else to bring together variant readings. Ingram and Swaim's classic of the concordance-maker's art is 'based on the texts of Milton's poems that were published in his lifetime, on certain authoritative manuscripts of the same period, and in a few instances on later-seventeenth-century texts as well'[4]. This brings together major textual variants, even if they were deleted in the manuscripts where they occur. Thus, under the word 'Go', we find:

> (As I will give you when we go) you may   Mask 648
> Trinity ms 'when we goe' <- 'on the way'
> <- 'as wee goe'
> Bridgewater ms 'goe'
> 1637 'goe'[5]

Such a wealth of information means the concordance can be used to answer questions which simpler concordances cannot help with. The concordance-maker must decide whether such advantages merit the multiplied task of preparing not only the base text, but also the appropriate alternatives.

Once the text has been decided on, it must be inserted into the computer, either typed up in the traditional fashion onto punch-cards or else keyed in directly onto disk or magnetic tape. Optical scanners are becoming more easily available to academic researchers and greatly facilitate input of material that is available in modern printed editions. In North America, the United Kingdom and in many other countries, optical scanners must now be regarded as the most cost-effective option where a large amount of material is to be entered and where it is available in a satisfactory modern printed version. The text, which now exists as a computer 'file', must now be proof-read. It is 'listed' (that is, typed out again by the machine), and the listing then read against the original to check for typing errors. Proof reading is still necessary with the current generation of optical scanners. The checking can be assisted by the generation of a 'word-list', a common facility of concordance pro-grams. This is a list of all the words in the text. Probable typing errors can be picked out in the version on file, located and, if necessary, corrected using the computer's 'editor'[6].

Next the version stored on file must have inserted into it the references which are to be included in the concordance to identify the location of each incidence in the base text. Such references may be quite complicated. Suppose, for example, we are preparing a corpus of dramatic literature, and wish the concordance to give, for each occurrence of every word, the play, the act, the scene, the line, and the character who speaks it. It is necessary to insert appropriate markers into the text on file. At the start of each play, each act, each scene, each speech, and perhaps—depending on the program used—each line, a reference must be introduced and it must be in some way distinguished from the raw text so it is not itself processed in the concordance-making run. Suppose, for example, we are producing a concordance to Shakespeare using COCOA, or, for that matter, using OCP, which will operate on texts prepared for COCOA. The text of *A Midsummer-Night's Dream* may be prefaced with

<P MND> <B 1> <S 1> <C THESEUS>

that is, *A Midsummer-Night's Dream*, act 1 (arbitrarily selecting 'B' to mean 'act'), scene 1, line 1, Theseus is the character who is speaking[7]. Seven lines into the text, we would need to insert a new character reference, <C HIPPOLYTA>; five lines further, then <C THESEUS> again. Before the second scene, we would need a new scene reference, <S 2> <L 1>, and so on. Of course, the procedure can be altered, if the concordance-maker so wishes, so that references are inserted into the text at the same time as it is entered into the machine, rather than edited in later.

We may want some material to appear in the contexts for items in the concordance but without itself being processed. For example, in *A Midsummer-Night's Dream*, I,i,16, we find the scene direction 'Exit Philostrate', We may wish this to be printed in the contexts for the words occurring around it, but perhaps would not want it to be sorted and reproduced under the headwords 'EXIT' and 'PHILOS-TRATE'. This need present no major difficulty. For example, if we were using COCOA, we would insert double round brackets around the scene direction.

There are strong arguments in favour of retaining the line-divisions of the base text. It makes line-references back to the base text much more coherent, facilitates proof-reading of the listing of the computer text against the original, and, when the project is over and the computer text archived, it leaves it in a state likely to be most useful to most who may wish to use it. If data preparation is done on punched cards, then the obvious measure is to use one card for each line of text. If a line spills over one card, it is possible to tell the machine to regard the second card, not as a new line, but as a continuation of the first.

Determining the Context

A concordance lists each incidence of the words within the text under an appropriate headword. Each incidence cited is embedded in a context and accompanied by a reference to its location in the base text. Determining the context is perhaps the greatest challenge to the concordance-maker's art. There are four basic strategies which may be adopted:

1. The machine is instructed to print a specified number of characters around the 'keyword' (i.e. the word being recorded). This produces a serviceable but shoddy output (Figure 6.1). Words in the contexts are truncated. Had OCP been used rather than COCOA, it would have been possible to specify that the context should delete any incomplete words. Nor do the contexts contain syntactically coherent units. However, perhaps for most purposes and, especially if it is to be used as a private research tool, such a concordance is quite adequate.

2. The concordance offers each keyword reproduced in the context of the line of text in which it occurs. This is a particularly popular formula in the processing of poetry where it generally works satisfactorily. As in the first strategy the context is determined entirely by the machine without editorial involvement, so it is a quick procedure. However, in poetry, at least in the English tradition, there is often a fair agreement between the ends of lines and syntactical units. Clauses often terminate at the end of lines. Consequently this strategy often produces a high proportion of contexts which are syntactically coherent, though, of course, it varies from poet to poet and period to period. Some of its disadvantages can be seen if the generally excellent *Concordance to Milton's English Poetry*, edited by Ingram and Swaim, is compared with its hand-compiled predecessor, which was edited by John Bradshaw in the late nineteenth century[8]. As can be seen from a comparison of Figures 6.2 and 6.3, the entries for 'Henceforth', the contexts in the older concordance, hand-selected by Bradshaw from the base text, sometimes provide a better guide to how the word is used than the machine-compiled concordance, which simply gives the line of occurrence, even though this may separate the keyword from the rest of the clause in which it occurs. Note especially the entries for *Paradise Lost*, III, 414, IV, 378, V, 881.

```
           9    BEEN

19  MAN, THAT ANY THING WHICH IS KNOWN OF LATE TO HAVE  BEEN REPRESENTED TO YOU: AND LASTLY, HAVING AMONG YO
73  P HE SHOULD TEACH ANY OTHER DOCTRINE, THAN WHAT HAD  BEEN BY SAINT PAUL PREACHED TO HIS GALATHIANS, WERE
100 P VERY THING, WHICH FROM SIMON MAGUS DOWNWARD, HATH  BEEN OBSERVED IN ALL HERETICS, CALLING THEMSELVES TH
111 R ALLEGIANCE TO HIM, TO WHOM THEY HAVE OFTEN BY LAW  BEEN REQUIRED TO SWEAR IT, A SECOND PRINCIPLE WHICH
127 SUCH APPEALS BY ORDEAL, OR DUEL, AND THE LIKE, HAVE  BEEN JUSTLY CENSURED AS UNCHRISTIAN AND BARBAROUS, N
45  EIR LAWFUL KING HEZEKIAH, THAT HIS MASTERS ARMS HAD  BEEN INVINCIBLE, THREE WHETHER THAT SADDEST FATE OF
59   THAT MOST COMMONLY THE PROSPERITY OF ARMS HATH NOT  BEEN THE LOT OF THE MOST RIGHTEOUS, BUT THAT EITHER
72  ) I CANNOT BUT TAKE NOTICE OF ONE PLACE, WHICH HATH  BEEN PRODUCED TO THAT PURPOSE, ONE PETER TWO THIRTEE
98  PON YOU, TO BE EATEN UP WITH WORMS, (AFTER YOU HAVE  BEEN THUS DEFIED) NOW, THAT YOU HAVE NO OTHER VISIB


           2    BEFORE

17  MORE IMMEDIATELY TO YOUR WELFARE, AND HONOUR, BOTH  BEFORE GOD AND MAN, THAN ANY THING WHICH IS KNOWN OF
47  ND EXAMINED REGULARLY, WHETHER IT BE OF GOD, OR NO?  BEFORE THE SUBJECT-MATTER OF SUCH REVELATION BE BELI


           1    BEGAT

31  AND THE GREEK CHURCH THE OTHER, AND THAT DIFFERENCE  BEGAT A WAR BETWIXT THEM, ITIS CLEAR THAT THE TURKS


           1    BEGET

95  OD, OR THE LYCAONIANS TO PAUL AND BARNABUS, IF THEY  BEGET NOT IN YOU A JUST INDIGNATION WITH THE LATTER,


           1    BEGINNING

25  TEEN FIFTEEN THE SAME WITH (THE WHOLE WORLD) IN THE  BEGINNING OF THAT VERSE, AND (ALL NATIONS) IN THE PA


           2    BEING

77  ER THE TITLE OF THE APOCALYPSE OR REVELATION (WHICH  BEING THE LAST WRITING WHICH IS KNOWN TO BE DICTATED
85  NCE TO BRING UP WITH HIM A LOST WORLD: AND THREE OF  BEING A LINEAMENT OF THE FACE OF DIVINE GOODNESS, IN
```

Fig. 6.1 Section of a simple concordance generated by COCOA

Henceforth
  Consult how we may henceforth most offend   *Par Lost 1.187*
    ms 'hence forth'
  Henceforth his might we know, and know our own   *Par Lost 1.643*
    ms 'Hence forth'
  Henceforth, and never shall my Harp thy praise   *Par Lost 3.414*
  Henceforth; my dwelling haply may not please   *Par Lost 4.378*
  Henceforth an individual solace dear;   *Par Lost 4.486*
  And seale thee so, as henceforth not to scorne   *Par Lost 4.966*
  Taste this, and be henceforth among the Gods   *Par Lost 5.77*
  Both of thy crime and punishment: henceforth   *Par Lost 5.881*
  Open, and henceforth oft; for God will deigne   *Par Lost 7.569*
  Created; but henceforth my early care,   *Par Lost 9.799*
  Henceforth of God or Angel, earst with joy   *Par Lost 9.1081*
  Let none henceforth seek needless cause to approve   *Par Lost 9.1140*
  And henceforth Monarchie with thee divide   *Par Lost 10.379*
  Henceforth; least that too heav'nly form, pretended   *Par Lost 10.872*
  I never from thy side henceforth to stray   *Par Lost 11.176*
  Henceforth I flie not Death, nor would prolong   *Par Lost 11.547*
  Henceforth to be foretold what shall befall   *Par Lost 11.771*
  Henceforth what is to com I will relate   *Par Lost 12.11*
  Henceforth I learne, that to obey is best   *Par Lost 12.561*
  And high prediction, henceforth I expose   *Par Reg 1.142*
  The Gentiles; henceforth Oracles are ceast   *Par Reg 1.456*
  And sends his Spirit of Truth henceforth to dwell   *Par Reg 1.462*
  He never more henceforth will dare set foot   *Par Reg 4.610*
  Henceforth, not too much disapprove my own.   *Samson 970*
  Hence forth thou art the Genius of the shore.   *Lycidas 183*
    Trinity as 'henceforth'
    1638 'Henceforth'

**Fig. 6.2 Section from *A Concordance to Milton's English Poetry*,
edited by W. Ingram and K. Swain.**

Henceforth. - P.L. 1, 187. how we may *h.*
  P.L. 1, 643. *h.* his might we know
  P.L. 3, 414. the copious matter of my song *h.*
  P.L. 4, 378. you must dwell or you with me *h.*
  P.L. 4, 486. to have thee by my side *h.*
  P.L. 4, 966. as *h.* not to scorn the facile gates
  P.L. 5, 77. be *h.* among the gods thyself
  P.L. 5, 881. *h.* no more be troubled how
  P.L. 7, 569. open and *h.* oft
  P.L. 9, 799. *h.* my early care not without song
  P.L. 9, 1081. how shall I behold the face *h.*
  P.L. 9, 1140. let none *h.* seek needless cause
  P.L. 10, 379. *h.* monarchy with the divide
  P.L. 10, 872. to warn all creatures from thee *h.*
  P.L. 11, 176. I never from thy side *h.* to stray
  P.L. 11, 547. *h.* I fly not death nor would
  P.L. 11, 771. let no man seek *h.* to be foretold
  P.L. 12, 11. *h.* what is to come I will relate
  P.L. 12, 561. *h.* I learn that to obey is best

P.R. 1, 142. *h.* I expose
P.R. 1, 456. *h.* oracles are ceased
P.R. 1, 462. Spirit of Truth *h.* to dwell
P.R. 4, 610. be never more *h.* will dare
S.A. 970. I desist *h.*
L. 183. *h.* thou art the genius of the shore

Fig 6.3 Section from *A Concordance to the Poetical Works of John Milton,* **edited by J. Bradshaw**

3. It is a fairly simple procedure to instruct the machine to extract, as a context, the words, around the key word, which are bracketed by punctuation. The machine, having located the key word, can work backwards to the previous comma, full stop, and so on, and forwards to the next one, and so isolate, most probably, a sentence or clause as the context for the key word. Such a facility is available in both COCOA and OCP. Of course, the context produced may be unwieldy, and so it may be useful to combine this instruction with another, setting an upper limit to the amount of context extracted and printed. A much subtler manoeuvre was incorporated into the program adopted by Spevack in his concordance to Shakespeare [9]. Whenever punctuation appeared within a context derived using a strategy similar to 1, the machine, reading first right and then left from the key word, was instructed:

> to stop as soon as a full stop was reached in one direction and then fill in the remaining blanks in the other direction until another full stop was found or there was no space left [10]. If, however, a half stop were encountered first, the computer would stop temporarily, reverse direction, and continue beyond the first encountered half stop until such a point as it found another signal to stop completely or reverse direction again... analysis... showed that quite often the verse line itself operates much like a syntactical unit. Consequently, the verse line ending without punctuation was, at the outset at least, considered a half stop, and the combination of punctuation mark and verse line ending was considered a full stop [11].

The result, for the most part, is a concordance characterised by the coherence and usefulness of its contexts. Figure 6.4 offers a sample. Howard-Hill, though generally sceptical about the sensitivity of algorithms to 'complex syntactical situations', commends Spevack's system to concordance-makers 'who do not intend to use other editorial methods [12]'. Two caveats, however, must be given. Such a system assumes a degree of coherence in the punctuation of the base text which may not be justified, especially if it is of a work from the early modern period or earlier [13]. Secondly, this procedure is not an option in either COCOA or OCP, and may well require a considerable degree of programming expertise.

4. *Post-editing*. The most satisfactory but also the most arduous strategy, which requires the concordance-maker to review each context and assess its adequacy. Probably the easiest way to do this is to store the output of the concordance-generating program as computer files, which can then be edited using the computer's editor. It is possible to have the program select a larger context around the key word than could be accommodated in the final form, and this can then be edited down by the simple deletion of whatever is judged redundant. Should a context still seem inadequate, it could be supplemented from the base text. Such a procedure requires, for even a modest corpus, the repetition many thousands of times of the same minute acts of judgement and computer manipulation, and will require long periods of computer connect time.

Orthographical Irregularity, Lemmatisation and Homograph Distinction

Unless instructed otherwise, the machine generates a concordance which separates the various inflected forms of the same lexical word (listing 'Dog' separately from 'Dogs', 'Man' from 'Men', 'Go' from 'Went') and examples of the same word spelt differently ('Intire' from 'Entire', and so on). At the same time, homographs will be sorted together. This may not seem satisfactory to the concordance-maker, to whom there are four strategies open:

1. Pre-editing
The insertion into the text at the preparation stage of markers which will distinguish homographs. For example, if a text contained both 'Lead' (OED sb.1) and 'Lead' (OED sb.2), the former could be entered, whenever it occurs in the text, as 'lead1', the latter as

'lead2', and they could thus be sorted separately when the concordance is generated. It will be obvious that such pre-editing may be quite arduous if many homographs are to be distinguished.

2. Programming

The machine can be instructed to sort together specified groups of inflected or orthographically distinct words. Such an option is available both in COCOA and OCP, but, once more, preparing the program to accommodate all the variations of a long text presents a messy and time-consuming chore, especially if spelling is very irregular. It is an easy matter, however, to group together some of the commoner problematic words, like, for example, the parts of the verb 'Be'.

3. Post-editing

After the concordance has been generated, it can be listed into computer files and then, by means of the insertion, deletion, and perhaps, block-movement facilities of the computer's editor, items may be moved together which, because of inflection or orthographical irregularity, would otherwise be printed apart, and homographs which are sorted together can be separated.

4. Cross-referencing and Annotation

Insertion into the concordance, alongside headwords, of notes drawing the user's attention to alternative forms and closely connected words which he may wish to look up. This policy has been adopted, with robust good sense, by workers on the Cornell University series of concordances. The concordance to the poetry of Swift, for example, includes cross-references 'when both the modern spelling and the old spelling variant exist and, in the concordance text, are farther apart than half a page (or more than about fifty lines apart), and when something other than the use of the apostrophe is responsible for the variant [14]. To alert users, Spevack marks with asterisks headwords under which homographs are brought together. Such annotation can, of course, be conveniently done using the computer's editor if the concordance, once it is generated, is held on file.

Limiting Output

In the case of the private concordance, the arguments against exhaustiveness are relatively weak. The expense of providing references and contexts for every incidence of every word within the corpus is no more than the cost of the computer time and the line-printer paper for the output. However, such lavishness is rarely possible in a public concordance, and it is customary to omit high-frequency words such as definite and indefinite articles, some conjunctions, some prepositions, some auxiliary words and parts of the verb 'Be'. A word list of the vocabulary in order of frequency can be generated and a cut-off point decided ('no words occurring more than 300 times', or whatever). The program is then adjusted to exclude these high-frequency words. If we were using COCOA, for example, our card 12 could read:

C<E5P3L3>100#100 C(A Z/1 300)

The phrase 1 300 means that we require a concordance of only words occurring between one and three hundred times in the text to be processed. To do this using OCP, we must include, in the *ACTION section, the command PICK FREQUENCY LT 301.

Sometimes, the concordance-maker may decide that it is appropriate to include some high-frequency words. The Cornell concordance to Blake, for example, excludes 'Do' and 'Did', but retains 'Doth' and 'Dost', 'partly on the ground that these uncommon forms of common words are of stylistic interest [15]'.

To retain some frequently occurring words while excluding others requires a slightly different programming procedure. The commands must nominate exactly which words are to be excluded. For example, if we were using COCOA, card 12 could read:

<E5P3L3>100#100 C>A THE AND BUT<

Here the phrase >A THE AND BUT< means that the words mentioned are not to be accepted as headwords, but all other words are.[16] For OCP, the *ACTION section must contain the command PICK EXCEPT WORDS 'A THE AND BUT'.

Of course, if common words are excluded from the full concordancing procedure, it may still be appropriate to insert information about their total incidence.

CONCORDANCE-RELATED OPERATIONS

If the base text has been transcribed into a computer-readable form, then some of the objectives customarily met by consulting a concordance can be realised without full concordance-generation. Concordance programs may be invoked to search texts in order to locate and give contexts for words that particularly interest the researcher. Suppose, for example, one wanted to locate terms expressing moral judgment in the novels of Jane Austen. If the material were available in a computer-readable form, it would be a simple matter to nominate the relevant words and ask the machine to locate them in the text and to produce a concordance just for those terms. Of course, it may be felt that the researcher would not know which words to look at if he did not know which words occurred in the text, so the first stage may be to command the machine to produce a word list. This can be fairly quickly examined and any words of possible interest nominated for closer scrutiny. Were we using COCOA, for example, card 12 would need to include a phrase such as C<GOOD BAD BETTER WORSE BEST WORST>. This would produce a concordance just for those words. Using OCP we would specify in the *ACTION section PICK 'GOOD BAD BETTER WORSE BEST WORST'. Were appropriate texts for comparison also available, they too, could be interrogated using the same commands from the concordance program, and contrasts and analogies drawn.

Particular forms of words may be searched out and considered. Does, say, Milton favour present participles? Since such participles in English always have the suffix '-ing', we can command the machine to search out and provide contexts for items with this ending. In COCOA, this would require, in card 12, the phrase C<=ING>. In OCP, the *ACTION section needs the command PICK WORDS '*ING'. Of course, such a concordance would include words like 'Darkling' and 'Darling', nor would it distinguish participles from verbal nouns, such as 'Bearing' (*Samson Agonistes*, line 665). Such distinctions can be made by the researcher himself as he works through the selective concordance he has generated.

The capabilities of a powerful concordance program can be adapted to a considerable range of purposes. It is possible to make a rudimentary appraisal of collocation. Words may be nominated for selection and their contexts can be scruntinised. We could, for example, call up colour-words from Milton's poetry, together with a generous amount of context from which to identify the nouns to which they relate. We could categorise the nouns—'abstract', 'concrete', 'animate', 'human', and so on—and assemble statistics. If the operation were repeated on a corpus of suitable texts selected for comparison, then eccentric or noteworthy aspects of Milton's practice would become apparent.

Again, both COCOA and OCP have the capacity to locate co-occurrences. For example, we may wish to identify co-occurrences in Milton's poetry of the terms 'God' and 'Man'. We could include on card 12 the phrase C<#GOD 5 MAN #MAN 5 GOD>, which would locate all the places where the terms occur with no more than five words separating them. Using OCP, we need specify in the *ACTION section PICK COLLOCATES 'GOD' UPTO5 'MAN'.

VOCABULARY STATISTICS

As concordances are generated it is usual for the number of examples found and listed to be stated alongside the headwords. This is an inevitable feature of output generated using COCOA, though OCP offers the option of suppressing the information. The possibilities for statistical analysis of such information are obvious. In OCP the researcher may request insertion alongside each headword not only of the absolute occurrence of the item, but also of its frequency as a percentage of all the words in the corpus processed. Spevack's concordance to Shakespeare gives the incidence of each word (as a percentage and in absolute terms) for the total vocabulary of Shakespeare, for the vocabulary of each play, and indeed, for the vocabulary of each character (see Figure 6.4). Such exhaustiveness could be realised through OCP.

We have spoken already of the computer's capacity by means of options available in concordance programs to generate word-lists, in which the vocabulary items within a text are listed, together with a count of their incidence. Such lists may be in alphabetical order, in reverse alphabetical order, in alphabetical order according to the endings of words, or, as we have seen, in ascending or descending order of frequency. Of course, lists can be derived for selected sections of a corpus, such as particular poems or the various characters within a play.

Both COCOA and OCP have the capability to generate frequency tables which show how many words occur once, twice, three times, and so on, and which express the findings in terms of all the words and all the different vocabulary items within the material processed (see Figure 6.5). The ratio of different vocabulary items to total number of words may well be an important aspect of style,

1. WITCH

23 SPEECHES	23 VERSE	0 PROSE	6 VERSE AND PROSE
61 LINES	61 VERSE	0 PROSE	2 SPLIT LINES
356 WORDS	350 VERSE	0 PROSE	227 DIFFERENT WORDS

3.643 PERCENT OF SPEECHES OF DRAMA
2.596 PERCENT OF LINES OF DRAMA
2.129 PERCENT OF WORDS OF DRAMA

WORD	FREQ	REL FR	V	P	ACT-SCENE-LINE REFERENCE
A*	7	2.000	7	0	1.03. 4 1.03. 8 1.03. 9 1.03. 9
					1.03. 21 1.03. 28 4.01.129
ABOUT	1	0.285	1	0	4.01. 4
AGAIN	2	0.571	2	0	1.01. 1 3.05. 36
AIR	1	0.285	1	0	4.01.129
ALEPPO	1	0.285	1	0	1.03. 7
ALL	5	1.426	5	0	1.03. 14 1.03. 16 1.03. 40 1.03. 69
					4.81.125
AMAZEDLY	1	0.285	1	0	4.01.126
AND	11	3.142	11	0	1.03. 5 1.03. 5 1.03. 5 1.03. 5
					1.03. 10 1.03. 15 1.03. 23 1.03. 65
					1.03. 69 4.01. 7 4.01.126
ANGERLY	1	0.285	1	0	3.05. 1
ANOTHER	1	0.285	1	0	4.01. 75
ANTIC	1	0.285	1	0	4.01.130
AROINT	1	0.285	1	0	1.03. 6
ART*	1	0.285	1	0	1.03. 12
AS*	2	0.571	2	0	1.03. 18 1.03. 29
AY*	1	0.285	1	0	4.01.125
BACK	1	0.285	1	0	3.05. 36
BANQUO	1	0.285	1	0	1.03. 69
BARK*	1	0.285	1	0	1.03. 24
BE*	4	1.142	4	0	1.03. 24 1.03. 25 3.05. 26 4.01. 75
BEEN	1	0.285	1	0	1.03. 1
BEST	1	0.285	1	0	4.01.128
BLOOD	1	0.285	1	0	4.01. 64
BLOW*	1	0.285	1	0	1.03. 15
BOIL	1	0.285	1	0	4.01. 9
BRINDED	1	0.285	1	0	4.01. 1
BUT	3	0.857	3	0	1.03. 8 4.01. 70 4.01.125
CANNOT	1	0.285	1	0	1.03. 24
CARD	1	0.285	1	0	1.03. 17
CAT	1	0.285	1	0	4.01. 1
CAULDRON	1	0.285	1	0	4.01. 4
CHARM	1	0.285	1	0	4.01.129
CHARMED	1	0.285	1	0	4.01. 9
CHEER	1	0.285	1	0	4.01.127
CHESTNUTS	1	0.285	1	0	1.03. 4
COLD	1	0.285	1	0	4.01. 6
COME	4	1.142	4	0	1.01. 8 1.03. 29 3.05. 36 4.01.127
COMMANDED	1	0.285	1	0	4.01. 75
CRIES	1	0.285	1	0	1.03. 6
DAY	1	0.285	1	0	1.03. 19
DAYS	1	0.285	1	0	4.01. 7
DELIGHTS	1	0.285	1	0	4.01.126
DID	2	0.571	2	0	1.03. 29 4.01.132
DO*	3	0.857	3	0	1.03. 10 1.03. 10 1.03. 10
DRAIN	1	0.285	1	0	1.03. 18
DRY	1	0.285	1	0	1.03. 18
DUTIES	1	0.285	1	0	4.01.132
DWINDLE	1	0.285	1	0	1.03. 23
EATEN	1	0.285	1	0	4.01. 64
ENTRAILS	1	0.285	1	0	4.01. 5
FARROW	1	0.285	1	0	4.01. 65
FIRST	2	0.571	2	0	4.01. 9 4.01. 76
FLAME	1	0.285	1	0	4.01. 67
FORBID	1	0.285	1	0	1.03. 21
FROM	3	0.857	3	0	4.01. 62 4.01. 63 4.01. 66
GIBBET	1	0.285	1	0	4.01. 63
GIVE	2	0.571	2	0	1.03. 5 4.01.129
GLAMIS	1	0.285	1	0	1.03. 48
GO	1	0.285	1	0	4.01. 4
GONE	1	0.285	1	0	1.03. 7
GOT*	1	0.285	1	0	4.01. 8
GRAYMALKIN	1	0.285	1	0	1.01. 8

Fig 6.4 Section from *A Complete and Systematic Concordance to the Works of Shakespeare*, **edited by M. Spevack.**

though, of course, a work in which allusion is made to many concepts may well draw upon a wider vocabulary than one in which an author considers a narrower subject in depth [17]. The sort of information which can be acquired using these options in concordance-generation programs is tractable to quite subtle application.

WORD COUNT	NUMBER SUCH	VOCAB TOTAL	WORD TOTAL	PERC. OF VOCAB.	PERC. OF WORDS
1	535	535	535	64.15	17.83
2	111	646	757	77.46	25.22
3	57	705	928	84.29	30.92
4	30	733	1048	87.89	34.92
5	11	744	1103	89.21	36.75
6	11	755	1169	90.53	38.95
7	15	770	1274	92.33	42.45
8	13	783	1378	93.88	45.92
9	4	787	1414	94.36	47.12
10	5	792	1464	94.96	48.78
11	5	797	1519	95.56	50.62
12	3	800	1555	95.92	51.82
13	1	801	1568	96.04	52.25
14	3	804	1610	96.40	53.65
15	1	805	1625	96.52	54.15
16	1	806	1641	96.64	54.68
18	3	809	1695	97.00	56.48
19	3	812	1752	97.36	58.38
20	1	813	1772	97.48	59.05
22	1	814	1794	97.60	59.78
27	2	816	1848	97.84	61.58
28	1	817	1876	97.96	62.51
29	1	818	1905	98.08	63.48
30	3	821	1995	98.44	66.68
33	1	822	2028	98.56	67.58
34	2	824	2096	98.80	69.84
43	1	825	2139	98.92	71.28
44	1	826	2183	99.04	72.74
48	1	827	2231	99.16	74.34
51	2	829	2333	99.40	77.74
81	1	830	2414	99.52	80.44
104	1	831	2518	99.64	83.91
134	1	832	1652	99.76	88.37
159	1	833	2811	99.88	93.67
190	1	834	3001	100.00	100.00

Fig 6.5 Word-frequency table generated by COCOA.

The OCP package has the capacity, before processing, to select a sample of any specified size, constructed by the random selection of words from the text under scrutiny. This may offer distinct advantages for some statistical procedures.

The tables and statistics produced by any of the operations described in this section count as 'different words' inflected forms and spelling variants, nor do they distinguish homographs. If the distortions this may entail seem unacceptable in the context of the particular research project, they can be mitigated by invoking the pre-editing and programming techniques discussed above.

INDEXES

Both concordance packages have the capacity to produce indexes, which are, in effect, concordances in which references are retained but without contexts. The structure and scope of such indexes can be controlled using techniques already discussed in the sections on concordance generation.

SUMMARY

This chapter discussed the function and design of concordances, the preparation of texts for concordance production, and the concordance software packages COCOA and OCP. It describes the generation of statistics relating to the incidence of words within a text and the use of the computer in the preparation of indexes.

NOTES
1. See above, Chapter 1
2. Godelieve L M Berry-Rogghe and T D Crawford, *COCOA Manual*, Didcot and Cardiff, 1973; Susan Hockey and Ian Marriott, *Oxford Concordance Program Version 1.0 Users' Manual*, Oxford, 1980.
3. See T H Howard-Hill, *Literary Concordances: a Guide to the Preparation of Manual and Computer Concordances*, Oxford, 1979, pp. 7–13. His account has been most useful in the preparation of this chapter.
4. William Ingram and Kathleen Swaim, *A Concordance to Milton's English Poetry*, Oxford, 1972, p. v.
5. Ibid., p. 207. 'Mask' refers to *Comus*; '1637' to the edition of 1637. The base edition for this poem is *The Poems of John Milton, Both English and Latin*, London, 1645. The reference to the Trinity MS means that the final version there is 'when we goe', which replaced 'on the way', which in turn replaced 'as wee goe'.

6. See above, Chapter 2.
7. The example is adapted from *COCOA Manual*, p. 5.
8. John Bradshaw, *A Concordance to the Poetical Works of John Milton*, 1894: London, 1965.
9. Martin Spevack, *A Complete and Systematic Concordance to the Works of Shakespeare*, Hildesheim, 1968–70.
10. He restricts contexts to a maximum of 48 'spaces' (characters, punctuation and blanks).
11. Spevack, IV, vii.
12. Op. cit., p. 37.
13. See Mindele Treip, *Milton's Punctuation and Changing English Usage, 1582–1676*, London, 1970, passim.
14. Michael Shinagel, *A Concordance to the Poems of Jonathan Swift*, Ithaca, 1972, p. x.
15. D V Erdman, *A Concordance to the Writings of William Blake*, Ithaca, 1967, p. xi.
16. Adapted from *COCOA Manual*, p. 26.
17. For a discussion of techniques appropriate to the stylistic appraisal of word frequencies, see Thomas N Corns, *The Development of Milton's Prose Style*, Oxford, 1982, p. 3.

7. STRUCTURED INFORMATION

INTRODUCTION

Most literary studies contain some sections of structured information, and some draw heavily on the management of databases.

Nearly every monograph-length study has a booklist or bibliography which is highly structured. Each entry will contain the same order of statements, and entries will be ordered with respect to each other according to structuring principles. Here are some typical examples:

> Heal, Ambrose, *The English writing-masters and their copybooks 1570–1800* (Cambridge, 1930).
> Waddington, Raymond B., 'Visual Rhetoric: Chapman and the Extended Poem', *English Literary Renaissance,* 13 (1983), 36–57.
> British Library, Harleian MS. 2107, Cheshire remonstrance 1642.

These were generated through the application of the following, entirely typical, structuring principles:

> /(<author's surname> <author's initials or given name> <title of work>) + (<number of volumes — if more than one> <place of publication> <date of publication>/ OR /<periodical name> <volume number> <date of publication> <page references>)/ OR /(<location of manuscript> <call number> <description of manuscript>)/

Entries will be formally ordered. Customarily, manuscripts will be sorted separately from printed material. Sometimes secondary material will be sorted separately from primary texts. Within each category, items will be listed alphabetically by author's surname, and, where an author is responsible for several items, these can be sorted further by some other principle, usually chronology. All the stages in the organisation of such a bibliography can be done using existing and available computer programs, which can order the statements within each entry, separate entries into broad categories (MS, primary, secondary, and so on), and order items within the categories according to alphabet and chronology. This chore, which has traditionally been performed by shuffling and sorting slips or index cards, can be executed fairly simply by machine if the researcher has kept his bibliographical information in a computer-readable form.

Of course, some projects call for the manipulation of more complex information. For example, a theatre-history study may require information not only about the authorship, title and date of publication of plays, but also about early performances, about actors, directors, designers, choreographers, and so on, about backers and reviews. Or else, perhaps, the researcher may wish to keep, among the other details, abstracts of each entry. So far we have spoken of databases in which the entries have been written documents of some kind, but the records may as easily contain information on people (such as authors, booksellers, reviews, printers) or institutions (paper factories, theatres, bookbinderies, and so on). Many studies produce masses of such information which may require not simply to be sorted and listed, but rather to be interrogated in complex ways to elicit conclusions. Whereas the advantages of using a computer for small-scale bibliographical listings are modest, in this sort of large-scale data-management operation it offers power, speed and flexibility.

ESTABLISHING THE DATABASE

A database contains two distinct dimensions: cases and fields. The individual items (books, people, institutions, or whatever), about which the researcher collects information, are termed 'cases'. A database which contains 500 such items will be said to contain '500 cases'. These would correspond, perhaps, to 500 cards in a conventional boxfile. Each case will contain a number of 'fields'. These are the categories of information contained on each case. They correspond to the pieces of information written on each index card in a box file. In a computer-based system it is usually very important that the same fields and the same ordering of fields are used for each case within the datebase. A database may be conceptualised as a matrix.

Figure 7.1 shows how such a matrix would be organized for a simple bibliography list.

case number	one	two	three	four	n
author	Griffith, Matthew	Hammond, Henry	Nedham, Marchamont	Stubbe, Henry	
title	The Fear of God and the King	To the Right Honourable the Lord Fairfax, and his Councell of Warre: The Humble Addresse	Interest Will Not Lie	An Essay in Defence of the Good Old Cause	
place of publication	London	London	London	London	
date of publication	1660	1649	1659	1659	
content	Royalist; argument for Restoration	Royalist; defence of Charles I	Republican; argument against Restoration	Republican; argument against Restoration	

FIELDS

CASES

Fig 7.1 Data-base Matrix for a Bibliography

Since for each case, information must be entered in the same order, it is often advisable to structure the collection of information appropriately. Rather than gather quantities of information, which you must then reorganise and standardise manually before entering it into the machine, you may prefer to collect your data in such a way that it can be entered without editorial involvement by yourself. Researchers will often find it useful to design for themselves a form of data-sheet that presents, for each case, the fields in which they are collecting information in the order in which the database is organised. Figure 7.2 is an example of a completed form for a large study of a seventeenth-century booksellers, prepared for management using the FAMULUS program. Different programs and different projects require different forms which the researcher may well have to design ad hoc, but the general advantages of this way of working are considerable.

As the project progresses, it will be possible to add further cases, update information, supplement information in cases already in the database, using facilities of the data-management program or the editing facilities of the computer.

MANIPULATING THE DATABASE

After the database is established, the user instructs the machine to load it into its working memory and to manipulate it in specified ways, and thereafter either to save and/or to print out the result of the manipulation. Different programs offer different facilities. All the following, however, are within the capabilities of some programs, and most are within the capabilities of all.

Sorting

Each case contains a number of fields. The researcher nominates the field according to which he wants his database sorted, and the machine reorganises the file in alphabetical or chronological order with respect to the nominated field. For example, the publishing historian whose work is used in Figure 7.1 may want a listing of his file by printer's name and by date. He would nominate the field PRIN (short for printer's name) and the field DATE and instruct the machine to sort. The result should be an alphabetical list of printers, with output organised by date, together with the other information in the file.

Listing

The computer can produce listing of the contents of the database file in which the ordering of the fields for each case has been altered and in which some fields may be omitted. The layout of the output can also be controlled and modified. This may be especially useful if the output is to be reproduced for publication or circulation.

STNO	241
WING	S 5777
READ	Bodleian, 4 March 81
AUTH	Strada, Famianus
TITLE	De Bello Belgico
LANG	English from Latin
TRAN	Stapylton, Sir Robert
PUBL	Moseley, Humphrey
ADDR	Princes Arms in St Pauls Churchyard
PRIN	None given
PLAC	London
DATE	1650
ANON	Prin
PREF	Epistle to Lord Pierrepoint
GENR	European History
LENG	259,000
FORM	Folio in fours
MISC	errata

Fig 7.2 Data-collection form for a data-management project

Searching

The researcher can nominate a particular field and instruct the computer to search that field in the cases of the database to locate specified words or combinations of words, and to list the cases where these terms are to be found.

Indexing

The computer can be instructed to provide an index to the words it finds within a specified field. This may then be used to consult a full listing of the database with much greater facility.

Miscellaneous facilities often exist to provide concordances to nominated fields, to merge a plurality of homologous databases, to provide separate listings of, for example, works by a plurality of authors, to limit the processing of the database to a specified portion of the file, and to eliminate from the searching, listing, indexing, and so forth, specified function words of high incidence.

EVALUATING A DATA-MANAGEMENT PROGRAM

A multiplicity of data-management programs have been produced and some can be run satisfactorily on microcomputers. The researcher may well have several alternatives open to him among the computer facilities to which he has access, and in choosing between mainframe and micro and between different programs on the same machine, the following considerations are pertinent:

How much information can be held about each case? How many fields per case? How much information within each field?

How many cases can be held in the database file?

Will the program perform all the operations the research project calls for?

How easy is it to use? How user-friendly? There may be some distinct advantages in a system that permits interactive interrogation of the database. The user may wish to consider the adoption of one of the database management systems

(DBMS), which permit access to, and manipulation of, databases.

There are now many DBMS to choose from, and the most popular allow relational operations. These include: Structured Query Language (SQL), Query-By-Example (QBE), the INteractive GRaphics and REtrieval System (INGRES), ORACLE, RAPPORT and the Personal Data System (PDS)

TRIAL RUNS

We most strongly recommend that researchers who are establishing databases perform, at an early stage, a pilot exercise using only a few cases. This will test whether their data-collection procedures work efficiently, may suggest modifications to their data-forms, if they are using them, and will confirm that the data-management program they have selected will meet their requirements.

The dangers of merely collecting data, without this trial stage, are no doubt obvious.

BIBLIOGRAPHIES: THE CASE OF THE EIGHTEENTH-CENTURY SHORT-TITLE CATALOGUE

Librarianship, bibliography and information science are in the process of rapid conversion to computer-based systems. Not only are individual libraries mounting their catalogues on computers and, indeed, conducting much of the business of library management by machine, but also on-line bibliographical services are coming into operation both within Britain and internationally. The DIALOG Information Retrieval Service, based in California, has mounted a large number of catalogues, mostly, it must be said, of interest to scientists and economists, which can be searched from remote sites in the UK (and, of course, in the USA). The BLAISE system (an acronym for British Library Automated Information Service) is a national network, available on a subscription basis, which carries, among other items, UK and US MARC (Machine Readable Catalogue) files, such as the British National Bibliography and the Catalogue to the Library of Congress. Perhaps the most interesting project and the most indicative of the potential usefulness of machine-mounted cataloging for the literary scholar, is the Eighteenth-century Short-title Catalogue, now available on-line through the BLAISE system[1].

This project is the compilation of a short-title catalogue of eighteenth-century English publications. The advantages of computer-basing such an operation will, by now, be apparent. Moreover, the database, kept on file, is open to remote interrrogation with great facility and high speed in ways that a conventional hard copy is not. To examine the resources of the catalogue, we entered the BLAISE system, from a remote site, via a Telecom line, and called up the appropriate catalogue. We conducted the following dialogue:

The computer asks:
SS1/C?   (=do you wish to make a
                                search statement, to initiate a
                                search or do you wish to make
                                some other command?)

User
(TW)CRUSOE   (=find all titles containing the
                                word 'Crusoe')

Computer
PROG:
SS(1) PSTG (90)   (=90 are listed)
SS2/C?   (=a second search statement or
                                some other command?)

User:
1 AND NOT DEFOE :   (=find all titles which satisfy
                                first search statement and are
                                not by Defoe)

Computer:
MM (DEFOE :) (2)   (=2 Defoes are listed as
                                authors:
1 2 DEFOE BN (AU)   1. B N Defoe has 2 titles,
2. 681 DEFOE D (AU)   2. D Defoe has 681 titles.
SPECIFY NUMBERS, ALL   Do you want one or both
OR NONE   or neither?)
User:
2   (=we are interested only in
                                excluding D. Defoe)

Computer:
PROG:
SS(2) PSTG (7)   (=7 are listed)
SS3/C?
User:
2 AND NOT LONDON (PL)   (=find all titles which satisfy
                                 second search statement and
                                 which were published outside
                                 London)
Computer:
PROG:
SS(3) PSTG (5)   (=5 are listed)

A further command to print a list produces the appropriate biblio-graphical information for each of the five titles. Thus, very quickly, we found all books mentioning 'Crusoe' in the title, by authors other than Daniel Defoe, and printed outside London.

Plainly, this is a powerful research tool of enormous potential. It can locate, almost instantaneously, works which could probably only be found after time-consuming perusal of printed catalogues. It can scrutinise the database year by year, town by town, library by library, format by format, subject by subject, and so on, sifting out works that satisfy specified criteria. Further, it can produce a mass of statistics about the work of the press of enormous significance for workers in the field.

SUMMARY

This chapter discusses the use of computers in the storage and manipulation of bibliographies and other projects in which large amounts of information is stored in a structured form. It describes database management and considers the characteristics of available software packages and gives an account of the Eighteenth-Century Short-Title Catalogue.

NOTES

1. For an account of how to use ESTC on line, see R C Alston, 'Searching ESTC Online. A brief guide', *Factotum* Newsletter of the XVIIIth century STC Occasional Paper 1, 1982. The origins and procedures of the catalogue are described in R C Alston and M J Jannetta, *Bibliograhy, Machine Readable Cataloguing and the ESTC*, London, 1978.

8. ENCODED INFORMATION

INTRODUCTION

We have discussed ways in which the computer can manipulate data presented to it in natural language, either as a literary text to be processed or else as a more structured data base. For some purposes, however, the machine will be much more useful if the data is numerically encoded. In general, this is the case in projects which involve a considerable element of quantification and which require some process of classification by the investigator before the machine can perform its contribution.

As we have argued, computers function badly at the grammatical and semantic levels. Presented with a passage of raw text, the machine is not fully capable of identifying nouns, verbs, adjectives, and so forth without persistent human prompting. Nor can it tell what the passage is saying. Perhaps, with laborious programming, it could recognise some words as belonging to nominated categories of special interest. It could be provided with a vocabulary of terms to do with, say, darkness or contemporary politics or medieval courtly love, and be required to signal the occurrence of these items when it encounters them in a given text. Of course, it will mistake some and miss others, but maybe, for some purposes, its performance will be adequate. This is a long way, however, from simulating the processes by which a competent reader decodes a complex text. What is said about darkness or politics or love is lost. Tone, ideology, sentiment are quite intractable to such analysis, nor can machines cope with operations which draw upon notions of literary tradition, such as mode and genre. No machine can tell the ironic from the non-ironic, epic from mock-heroic, first-person fictional narrative from autobiography, nor distinguish the innovative or the archaic from the conventional.

However, in many cases, operations which invoke these levels of response are tractable to computer-aiding if the investigator first performs those tasks of interpretation which the machine cannot reproduce.

PROCEDURES ADAPTED FROM THE SOCIAL SCIENCES

The student of literature may well have something to learn from the investigative procedures by which sociologists and social historians sometimes develop statements about discrete areas of human activity. Let us consider a typical example and see what elements of its method may be adaptable to our purposes.

Suppose we wanted to appraise the level of literacy of a certain period of English history[1]. It would be necessary to develop an acceptable criterion for what constitutes literacy. It would have to be a test of performance which occurs throughout the period in more or less the same circumstances. Let us agree to term people 'literate' if they can endorse documents by signing their names rather than making their marks. This has the advantage of presenting a fairly objective test, though the investigator will sometimes have to take into account the possibility that a person may be too sick or crippled to sign. How next to proceed? The researcher will assemble as many documents bearing signatures or marks as seem appropriate and will collect other pertinent information about the people concerned. In a proper study, of course, factors other than simple date will require investigation. Towards constructing explanations for the level of literacy the researcher will be looking for the co-occurrence of other circumstances, such as the geographical location, gender, and social class of the people signing or marking, as well as the date. To the sets of pertinent alternatives ('It is a signature/mark'; 'the person is male/female'; 'he or she lives in area a or b or c, etc.'; and so on), we give the name, 'variables'. To every case in our study several such variables will pertain. In some of the variables, classification will be straightforward. For example, in each case the gender of the person will be either known or not known, and, if known, it will obviously be classifiable as male or female. However, some categories will require different kinds of decision by the researcher. Categorising social class, for example, may require the researcher to make an act of judgement which draws upon his larger perception of society at the time involved. Once all the information has been assembled, the computer can sift and calculate to indicate the relationships between the variable of literacy and the other pertinent variables.

Many aspects of the study of literature invite a similar methodology. Where the researcher must describe and analyse a considerable corpus of material to which a number of variables pertain, then the sort of procedures employed in the study of literacy are probably worth consideration.

Suppose, for example, we were investigating the English periodical press in the period 1700–1760. There were close to nine hundred papers in press during these six decades[2], and we may wish to consider a number of aspects of each. Each aspect may be regarded as a variable, to be treated like the variables in the study of literacy. For each periodical we may want to look at its date of inception, duration, frequency and place of publication, the sort of subjects it dealt with, whether it contained hard news, how many articles it usually contained, whether it carried advertisements, whether it printed letters from readers, how much it cost, what its ideological orientation was, and so on. Each periodical will require classification according to its variables, and, as in the study of literacy, sometimes classification will be straightforwardly objective (such as, say, its average length), whereas other categories (such as its ideological orientation) will call for different kinds of judgement. By adapting some of the methods of the earlier study we could achieve not simply a superficial description of the corpus, but perhaps work towards a much more complex analysis, ranging over many considerations. A vast array of questions can be raised and hypotheses tested. Was a paper that carried advertisements characteristically cheaper than one that did not? Did it run for longer? Were letters more common in provincial than London publications? Were weeklies containing hard news longer than weeklies that did not? Did Whig papers appear more often than Tory papers in the 1710s? And so on. The potency and the flexibility is surely obvious—especially if we have transferred the task of sorting and manipulating and calculating these variables to the computer. It is not simply that we can look at the issues more quickly than we could, were we collecting information in a card-index. More important is the way in which we can interrogate our information time and time again, sifting it, selecting from it, controlling some variables the better to explore others, following any hunch or insight, and wringing from it connections, possible causal links, which a non-computer analysis would probably have missed.

Later in this chapter we shall work through all the stages of a similar investigation, a quantitative account of the Thomason Collection of English Civil War pamphlets held in the British Library.

ENCODING INFORMATION

For each case in our study we will have a series of variables which could, perhaps, be represented, in a structured sort of way, in natural language, much as we recorded data in the data-management programs discussed in chapter seven:

study number	date of inception	place of publication	ideology
one	1706	London & Bath	Whig
two	1710	London	uncommitted

It would be possible to program the computer to perform the sorting, sifting and calculation on data presented to it in this form, but such a procedure would necessitate wearisome data-preparation and be indefensibly prodigal of computer resources. Each piece of information may occupy as much space in the computer memory as the longest term or phrase occurring to describe that variable (though the recent trend, in user-friendly systems, is not to restrict field sizes). The longest alternative found for the variable 'place of publication' may be 'London, Bath and Norwich'—twenty-four characters, including spaces. Not only would so much space be required wherever this alternative appeared, but also the computer most probably would have to leave the same amount of space in its memory for all the other alternative terms for this variable. It is customary, therefore, in this sort of computer application, to encode data for more efficient storage and manipulation.

Encoding is much simpler than it may seem. As in data-management programs, the computer is being asked to operate on a two-dimensional matrix, this time of the structure:

	variable 1	variable 2	variable 3
case 1			
case 2			
case 3			

Encoding information requires two stages. First, the machine must be told the structure of information in the data base—how to recognise where one case begins and another ends; the space allocated for each case; the location of the variables within the case allocation. Thus, for example, we may instruct the machine that each new line of the data file will initiate a new case, that in each line characters 1–4 will contain the case number, characters 5–6 will contain information on variable1, 7–8 on variable2, 9–10 on variable3, and so on. The second stage is to attribute a numerical code to each category that pertains to each variable. We construct a kind of code-book for all the alternatives for all the variables. Thus, for the variable 'place of publication', we may produce the following coding:

London	1
Bath	2
Norwich	3
Bristol	4
York	5
London and Bath	6
London, Bath and Norwich	7
other	8
not known	9

Suppose we had nominated that, in every line of our data file, the twelfth character would record information about place of publication, we could, for every periodical in our study determine the appropriate code number and insert it into its slot. Not all material, of course, will require this kind of encoding. Items like dates can be entered unconverted. Again, some variables consist not of discrete categories but of continuous ranges. Thus, while a periodical may be printed in town a, or town b, or town c (represented, say, by 1 or 2 or 3 in the code-book), it will not be printed at some nameless intermediary location (to be represented, perhaps, by 1.3 or 2.4 in the coding): these categories are discrete and non-continuous. But its length may be anything from about 500 words to 1000 words, and this continuous variable can often usefully be represented as such by the actual measurement.

PACKAGES, SAMPLING AND STATISTICS

Very many social scientists and researchers in related fields have undertaken projects using the methodology we have adapted, and so a number of packages of ready-prepared programs have been made widely available and some are supported by excellent documentation. A number, such as the SCSS Conversational System, permit interactive computing. We shall shortly be considering more fully an example of a project which uses perhaps the best known and most widely implemented package, the SPSS System (the Statistical Package for the Social Sciences)[3].

We have been discussing projects which attempt a general appraisal of large corpora of material. Sometimes, of course, the volume of texts will be too great to permit exhaustive consideration, but the problem can often be resolved satisfactorily by sampling. The researcher constructs a selection which properly represents the whole volume of cases. Thus, if we construct a properly representative sample of English periodicals from the period 1700–1760, then in the sample we will find approximately the same proportion of items printed in Bath, or in 1710, or of a Tory orientation as we would find in the whole volume of periodicals. Of course, the sort of precision we are aspiring to is vitiated unless the sampling operation is done properly. It will not do simply to include items because they are particularly interesting or easily available or because we believe that we intuit them to be representative. Rather, the sample must be constructed objectively, by using some random sampling technique. Introductory textbooks on statistics explain the appropriate methods[4]. Similarly, these works describe how to apply appropriate tests for the level of the significance of the co-occurrence of variables and other statistical procedures which may be pertinent in the analysis of our data and which the machine can perform for us.

AN EXAMPLE OF AN SPSS APPLICATION

Objectives

This is an account of the stages in setting up and executing a computer-based exercise in literacy and publishing history, 'Publication and Politics, 1640–1661: on SPSS—based Account of the Thomason Collection of Civil War Tracts', by T. N. Corns, *Literary and Linguistic Computing*, 1 (Sept 1986). Its purpose is to describe the non-periodical items in the collection of publications made by the London bookseller, George Thomason, during the period 1640–1661, held in the British Museum Library and recently made available on microfilm.

Sampling

Such a large opus, some 15000 items, could not practically be worked through by a solitary researcher in less than a lifetime, but so large a 'population' invites a sampling procedure. It was decided to construct a one-in-twenty sample by starting at a randomly selected point in the first twenty items in the catalogue to the collection and then selecting every twentieth item for analysis. This produced a data base of about 750 cases.

Data Collection and Preparation

The study assembled information about the circumstances of publication, about formal considerations, such as length and genre, about subject matter, and about ideology. Through a priori assumptions, tested and modified in pilot studies, the variables were decided upon and the categories determined. In all, 32 variables were chosen. Some procedure had to be developed for assembling 32 pieces of information about 750+ cases (over 24000 items of data in all) and then for transcribing this mass into a computer-readable form. For such an operation, a form or questionnaire is invaluable, and one was designed for the purpose (Figure 8.1). SPSS had been chosen, and it had been decided that the data for each case could be accommodated in one line of the data file. The questionnaire prompts the researcher with a brief description of each variable (together with the place in the line of the data file where it is to be stored). Next to each variable is printed a list of the alternative categories which pertain, together with their numerical code. Two points about questionnaire design should be noted. It is easier to manipulate a large stack of questionnaires if they can be printed on one side of a single sheet, though this, of course, is not essential. Secondly, it is quicker to transcribe the questionnaires if they contain columns into which the researcher can insert the appropriate code. After the texts have been analysed and the forms completed, each case can be represented by a single line of numbers. The lines that make up our data are then transferred to an appropriately formatted data-sheet and input to the machine. Of course, there may be errors, and the file must be listed, proof-read and corrected. Key—to disk entry is an alternative.

MS

1-4	Study no............		
	Thomason No..........		

6	Title	Biblical	1
		Classical	2
		Proverbial	3
		Other	4
		None	5

8	Printed 1	MS 2

10	Author	Name given	1
		Initials given	2
		Pseudonym	3
		Anon.	4

12		Individ. author	1
		Collaboration	2
		Collect. author	3
		N.K.	4

14	Printer	Name given	1
		Initials given	2
		Pseudonym	3
		Anon.	4
		N.A.	5

16	Publisher	Name given	1
		Initials given	2
		Pseudonym	3
		Anon.	4
		N.A.	5

18	Place of publication (as stated)	London	1
		Oxford	2
		Amsterdam	3
		Other	4
		Not given	5
		N.A.	6

20	Date ....
-23	

25	Motto	yes 1	no 2	N.A. 3

27	Cartoon/illustration/fp.			
		yes 1	no 2	N.A. 3

29	Errata	yes 1	no 2	N.A. 3

31	Licensed	yes 1	no 2	N.A. 3

33	Is it avowedly a reply? yes 1	no 2

35	Occurs in one edition	1
	two editions	2
	More than two	3
	N.A.	4

37	Format	Broadside	1
		Single Sheet	2
		Folio	3
		Quarto	4
		Other	5
		N.A.	6

39	Length
-43	

45	Language	English	1
		Dutch	2
		French	3
		Latin	4
		Other	5

47	Medium	Prose	1
		Verse	2
		Verse & Prose	3

49	Printed by order of	Charles I	1
		Charles II	2
		Parl/Prot	3
		None	4
		N.A.	5

51	Subgenre	Sermon (del)	1
-52		Sermon (not del)	2
		Prayer	3
		Treatise (theol)	4
		Exegesis	5
		History	6
		Declaration	7
		Speech (del)	8
		Speech (not del)	9
		Elegy/f'ral orat	10
		Dem'd/petition	11
		Lampoon	12
		Disp'tion/d'log	13
		Autobiog/apol	14
		Allegory	15
		Ballad	16
		Letter (auth)	17
		Letter (unauth)	18
		Newsb'k/rep't	19
		Plain expos'n	20
		Other	21

Area of reference
Main 1 secondary 2 periph/N.A. 3

54	Prel doctrine & disc	1 2 3
56	Sectarianism	1 2 3
58	Other theol issues	1 2 3
60	Civil gov't of Chl (pre'40)	1 2 3
62	Civil gov't of Chl (post'40)	1 2 3
64	republic'ism/regicide	1 2 3
66	Civil gov't-interregnum	1 2 3
68	Restoration	1 2 3
70	Other political issues	1 2 3
72	Non-controversial	1 2 3

74	Ideological orientation	Conservative	1
		Mediating	2
		Radical	3
		Ultraradical	4
		Uncommitted	5
		Uncertain	6

76	Literal 1	Ironic/parodic 2

Fig 8.1 Data collection form for a statistical project

```
RUN NAME        THOMASON PROJECT 81
FILE NAME       FILE 1, FIRST LIST FOR THOMASON PROJECT
DATA LIST       FIXED/1 STUDY NO 1-4, TITLE 6,
                PRINTMS 8, AUTHNAME 10, AUTHSTAT 12,
                PRINTER 14, PUBLSHER 16, PLACE 18, DATE 20-23, MOTTO 25,
                CARTILFP 27, ERRATA 29, LICENSED 31, REPLY 33, FORMAT 37,
                LENGTH 39-43, LANGUAGE 45, MEDIUM 47, BYORDER 49, SUBGENRE 51-52,
                PRELDD 54, SECTISM 56, OTHEOL 58, CGCHIB 62,
                REPUBLIC 64, CGINTER 66, RESTORAT 68, OPOL 70, NONCONT 72,
                IDEOLOGY 74, MODE 76
INPUT MEDIUM    DISK
N OF CASES      753
MISSING VALUES  PRINTMS TO PLACE MOTTO TO FORMAT, LANGUAGE TO BYORDER, PRELDD TO
                MODE (0)/DATE (0000)/LENGTH (00000)/
VAR LABLES      STUDYNO,STUDY NUMBER/ TITLE,ORIGIN OF TITLE/
                PRINTMS,PRINTED OR MANUSCRIPT?/
                AUTHSTAT,STATUS OF AUTHORSHIP/
                PUBLSHER,PUBLISHER/
                PLACE,PLACE OF PUBLICATION AS STATED/
                DATE,DATE OF PUBLICATION/
                MOTTO,IS THERE A MOTTO?/
                CARTILFP,IS THERE A C"R"T"N, ILL"ST"N OR F"P"CE?/
                ERRATA,IS THERE A LIST OF ERRATA?/
                LICENSED,IS IT STATED TO BE LICENSED?/
                REPLY,IS IT AVOWEDLY A REPLY?/
                LENGTH,LENGTH IN THOUSANDS OF WORDS/
                BYORDER,IS IT PRINTED BY ORDER?/
                PRELDD,PRELATICAL DOCTRINE AND DISCIPLINE/
                SECTISM,DISPUTES WITHIN PURITANISM/
                OTHEOL,OTHER THEOLOGICAL ISSUES/
                CGCHIA,CIVIL GOV"T OF CHARLES I BEFORE 1640/
                CGCHIB,CIVIL GOV"T OF CHARLES I AFTER 1640/
                REPUBLIC,CONSTITUTIONAL SHIFT OF 1649/
                CGINTER,CIVIL GOV"T OF INTERREGNUM/
                RESTORAT,RESTORATION/
                OPOL,OTHER POLITICAL ISSUES/NONCONT,NON-CONTROVERSIAL/
                IDEOLOGY,IDEOLOGICAL ORIENTATION/
VALUE LABELS    TITLE (1)BIBLICAL (2)CLASSICAL (3)PROVERBIAL (4)OTHER (5)NONE/
                PRINTS (1)PRINTED (2)MANUSCRIPT/
                AUTHNAME (1)NAME GIVEN (2)INITIALS GIVEN (3)PSEUDONYM (4)ANON,/
                AUTHSTAT (1)INDIVIDUAL AUTHOR (2)COLLABORATION (3)COLLECTIVE
                AUTHOR (4)NOT, KNOWN/
                PRINTER,PUBLSHER (1)NAME GIVEN (2)INITIALS GIVEN (3)PSEUDONYM
                (4)ANON, (5)NOT APPLICABLE/
                PLACE (1)LONDON (2)OXFORD (3)AMSTERDAM (4)OTHER (5)NOT GIVEN (6)
                NOT APPLICABLE/
                MOTTO TO LICENSED (1)YES (2)NO (3)NOT APPLICABLE/
                REPLY (1)YES (2)NO/
                FORMAT (1)BROADSIDE (2)SINGLE SHEET (3)FOLIO (4)QUARTO (5)OTHER
                (6)NOT APPLICABLE/
                LANGUAGE (1)ENGLISH (2)DUTCH (3)FRENCH (4)LATIN (5)OTHER/
                MEDIUM (1)PROSE (2)VERSE (3)VERSE AND PROSE/
                BYORDER (1)OF CHARLES I (2)OF CHARLES I (3)OF PARL OR PROTECTOR
                (4)NO-ONE (5)ANY OTHER AUTH"TY (6)NOT APPLICABLE/
                SUBGENRE (1)SERMON DELIVERED (2)SERMON NOT DELIVERED (3)PRAYER
                (4)TREATISE THEOLOGY (5)EXEGESIS (6)HISTORY (7)DECLARATION
                (8)SPEECH DELIVERED (9)SPEECH NOT DELIVERED (10)ELEGY OR F"RAL
                ORAT. (11)DEMAND OR PETITION (12)LAMPOON (13)DISPUT"N OR DIALOGUE
                (14)AUTOBIOG OR APOLOGY (15)ALLEGORY (16)BALLAD (17)LETTER
                AUTHORISED (18)LETTER UNAUTHORISED (19)NEWSBOOK OR REPORT
                (20)PLAIN EXPOSITION (21)OTHER/
                PRELDD TO NONCONT (1)MAIN AREA OF REF. (2)SEC"DARY AREA OF REF.
                (3)INSIGNIFICANT/
                IDEOLOGY (1)CONSERVATIVE (2)MEDIATING (3)RADICAL (4)ULTRARADICAL
                (5)UNCOMMITTED (6)UNCERTAIN/
                MODE 91)LITERAL (2)IRONIC OR PARODIC/
```

Fig 8.2 Defining an SPSS File

Defining the SPSS File

The machine, of course, needs guidance on how to interpret and manipulate the mass of numbers it receives. The internal structure and significance of the file must be defined. This is done through a series of defining commands (Figure 8.2). The file is given a name to be printed on future output through the command FILE NAME. Each variable is given a short title (up to 8 characters) and its location in the line specified by the command DATA LIST. The machine is instructed on how to respond to places in the data base where information is incomplete (MISSING VALUES). Fuller titles of the variables and the names of the encoded categories are specified, to be printed in future output (VAR LABELS and VALUE LABELS).

Descriptive Statistics

The data base can now be interrogated with facility to produce a general description of the corpus. Thus, for example, to discover the proportion of pamphlets bearing a motto on their title pages we need the command:

FREQUENCIES   INTEGER=MOTTO(0,3)

This produces the output illustrated in Figure 8.3.

FILE FILE1 (CREATION DATE = 19/02/81)   FIRST LIST FOR THOMASON PROJECT

MOTTO IS THERE A MOTTO?

CATEGORY LABEL	CODE	ABSOLUTE FREQUENCY	RELATIVE FREQUENCY (PERCENT)	ADJUSTED FREQUENCY (PERCENT)	CUMULATIVE ADJ FREQ (PERCENT)
YES	1	219	29.1	29.2	29.2
NO	2	528	70.1	70.3	99.5
NOT APPLICABLE	3	4	.5	.5	100.0
	0	2	.3	MISSING	
	TOTAL	753	100.0	100.0	

VALID CASES   751   MISSING CASES   2

Fig 8.3 An example of descriptive statistics using SPSS

Selecting From and Redefining the Data base;
Cross-tabulation

SPSS offers the facility of selecting for analysis from the data-base cases which satisfy specified criteria. Further, the categories assigned to each variable can be redefined to allow some to be merged. Such procedures are often useful in any statistical account and are sometimes essential for the next stage of analysis, which involves investigation of relationships of co-occurrence between two or more variables. A procedure which proves particularly useful in the context of this study is cross-tabulation analysis. Let us suppose we want to test a hypothesis, one of the hundreds which may be suggested and entertained in consideration of this data base. Let us consider the proposition that, in the context of the debate within puritanism, between, on the one side, Independents and radical sectaries, and on the other, Presbyterians and others who sought to control them, the former was less likely than the latter to include the imprimatur of a licenser in their publications. First, we must select from all the cases in the data base those which have this debate as their main area of reference. These are the items which have for the variable 'Sectarianism' (or 'Sectism' for short) the value 1. We command the machine:

SELECT IF   (SECTISM EQ 1)

('EQ' is short for 'equals'). There are two variables to be considered. 'LICENSED' (i.e., whether the publications carry the imprimatur of a licenser) offers three options—'yes' (1), 'no' (2), and 'not applicable' (3), the last pertinent in the case of manuscript material. This final category we shall exclude because it is not really relevant to the hypothesis under examination. Probably the easiest way to get rid of it is to amend the previous command, thus:

SELECT IF [(SECTISM EQ 1) AND   (LICENSED EQ 1
                                OR
                                LICENSED EQ 2)]

We need now to redefine the variable 'IDEOLOGY', which has 6 alternatives, 'conservative' (1) (in the context of this debate, denoting anti-Independent and anti-sectary, generally, though not exclusively, Presbyterian), 'mediating' (2), 'radical' (3) (here, Independents), 'ultra-radical' (4) (ranters, quakers, seekers, anabaptists, and the like), 'uncommitted' (5), and 'uncertain' (6). To test the hypothesis we want to exclude categories 2, 5, and 6, and combine categories 3 and 4. This we can probably do most easily by regrouping all the categories and then excluding the unwanted ones from further calculation. The newly formed categories can be appropriately labelled.

RECODE	IDEOLOGY(1=1) (3,4=2) (2,5,6=3)
MISSING VALUES	IDEOLOGY(3)
VALUE LABELS	IDEOLOGY(1)PRESB(2)IND,SECT

We are now ready to cross-tabulate the two variables which interest us. At the same time we shall have the machine perform a calculation to determine the statistical significance of the cross-tabulation, in this case, the familiar Chi-square test:

CROSSTABS	IDEOLOGY BY LICENSED
STATISTICS	1

The resulting table (Figure 8.4) shows a high level of correlation between Presbyterian ideology and the printing of imprimaturs, a useful fact for students of publishing history in that the imposition of licensing and the requirement of the imprimatur were Presbyterian measures designed to restrict the radicals: almost only Presbyterians obeyed the letter of their law.

Other Procedures

SPSS offers a very wide range of options in selecting and manipulating the data base, in performing calculations upon it, and in formatting the output. Our examples have been intended merely to illustrate some of the procedures which are useful in literary computing. Of course, this account in no way reduces the absolute necessity of using the appropriate SPSS manuals[5].

FILE FILE1 (CREATION DATE = 19/02/81) FIRST LIST FOR THOMASON PROJECT

★★★★★★★★★★ C R O S S T A B U L A T I O N ★★★★★★★★★★

IDEOLOGY IDEOLOGICAL ORIENTATION   BY LICENSED IS IT STATED TO BE LICENSED?

	LICENSED		
COUNT			
ROW PCT	YES	NO	ROW
COL PCT			TOTAL
TOT PCT	1.	2.	
IDEOLOGY			
1.	17	29	46
PRESB	37.0	63.0	62.3
	89.5	42.0	
	19.3	33.0	
2.	2	40	42
IND.SECT	4.8	95.2	47.7
	10.5	58.0	
	2.3	45.5	
COLUMN	19	69	88
TOTAL	21.6	78.4	100.0

CORRECTED CHI SQUARE = 11.60723 WITH 1 DEGREE OF FREEDOM. SIGNIFICANCE = .0007

NUMBER OF MISSING OBSERVATIONS = 6

Fig 8.4 An example of crosstabulation using SPSS

SUMMARY

This chapter describes the adaptation of techniques of the social sciences to the study of literature. We consider statistical procedures which may be appropriate in some areas of literary study and describe the design of such projects and their implementation using the computer. We give an example of a study using the software package SPSS.

NOTES

1. The example was suggested by John Cressy's admirable study, *Literacy and the Social Order: Reading and Writing in Tudor and Stuart England*, Cambridge, 1980.
2. Douglas Bond, in the Introduction to *Studies in the Early English Periodical*, ed. D Bond, Chapel Hill, 1957, pp. 3–4.
3. SPSS and SCSS are trademarks of SPSS Inc. of Chicago, Illinois, for its proprietary computer software. No materials describing such software may be produced or distributed without the written permission of SPSS Inc.
4. The student of literature may find the following particularly accessible: Anthony Kenny, *The Computation of Style*, Oxford, 1982; C B Williams, *Style and Vocabulary: Numerical Studies*, London, 1970; G Udny Yule, *The Statistical Study of Literary Vocabulary*, Cambridge, 1944.
5. Norman H Nie et al, *SPSS Statistical Package for the Social Sciences*, second edition, New York, 1975.

9. AUTHOR IDENTIFICATION AND CANONICAL INVESTIGATION

INTRODUCTION

The only aspects of computer application in the study of literature to secure a widespread attention which extends beyond the specialist interest of the academic disciplines involved are author identification and canonical investigation. Even now, the sort of studies that suggest a possible addition to the Shakespeare canon or dispute the authorial integrity of a book of the Bible may be reported widely. Armchair experts still muse that this or that ancient question of authorship may be resolved by 'running it through a computer'. The procedures involved are, of course, much more demanding than that facile phrase would indicate. Nevertheless, a satisfactory methodology has evolved for such projects and some major issues would seem to have been settled. However, three points must be stressed: this sort of enquiry calls for scrupulous attention to problems of method; it rests heavily on the application of statistical procedures which are often complex and sometimes controversial; and it requires the investment of considerable human resources in large-scale data-preparation processes[1].

THE NATURE OF THE PROBLEMS

The circumstances of composition surrounding most major literary texts from the modern period are generally transparent. The work bears the name of its author or authors, about whose claim various kinds of external evidence (diaries, letters to publishers, the testimony of friends and family, and so forth) provide ample corroboration. However, with some kinds of literature and in certain periods the circumstances of composition may be much less certain.

External evidence may be incomplete, misleading or contradictory. At this point, students of literature have sometimes turned to evidence that is internal to the problematic text. The problems to be resolved are fundamentally of four cognate structures:

> There may be reason to doubt the integrity of a work. The investigation seeks to show that the parts which make up a text are the product of one writer or of a plurality of writers.

> A work may be anonymous or pseudonymous. The investigation seeks to find a likely author from a multiplicity of possibles.

> External evidence suggests the work is the product of either A or B. Internal evidence is used to locate the text in the canon of one or other of the alternatives.

> External evidence indicates that work is a product of collaboration between A and B. Internal evidence effects the disintegration of the text, attributing segments appropriately.

A BASIC ASSUMPTION

Underlying the procedures developed for the use of internal evidence is a basic assumption that a writer's practices contain certain stable and distinctive elements which he cannot disguise and which others cannot simulate. This use of internal evidence is an exercise in detection, not aesthetic response. The features identified as characteristic may well have little to do with literary merit and with those qualities which make the work worth investigating in the first place. To use an analogy often cited in canonical studies, possible suspects are 'finger-printed', and the deed in question, the disputed text, scrutinised for clues.

VALIDITY AND VARIETY OF INTERNAL EVIDENCE

Scholars have used internal evidence soundly for several decades now, and certain criteria have emerged by which the validity of internal evidence is to be judged. A recent investigator offers the following useful list:

'Reliability may be achieved by using only items that are;

1. objective: unambiguously defined features, recognition of which is not a matter of opinion;
2. quantifiable (since differences between authors are in most features matters of greater or lesser frequency, not of invariable use or non-use).

Validity may be achieved by using only items that are;

3. likely (a priori) to correlate with authorship rather than with other factors (the other factors including transcription, compositorial practices, influences of genre or theatrical fashion, imitation, censorship, or common use in the period);
4. shown (a posteriori) to correlate with authorship rather than with such other factors, by means of comparisons which will tend to eliminate these factors'[2].

Over the years, many features have been regarded as characteristic of this or that author. Some, however, do not, as it has been observed, fully satisfy these criteria. It will not do to intuit the quality of Shakespeare's or Milton's hand in a text: there are good lines found in bad poets, and vice versa, nor will such intuition ever provide the strength of proof we require. Again, some features which have often been invoked, such as ideology or imagery, may be simulated or dissimulated, and may change with time or situation.

TYPES OF EVIDENCE AND THE ROLE OF THE COMPUTER

Of course, before committing themselves to so demanding an undertaking, researchers ought to consider fully the types of internal evidence adopted in previous investigations. Our study cannot appraise all the alternatives, but here are some of the more widely used.

Word Length and Sentence Length

Averages and related statistics have sometimes been thought worth calculating as distinctive features of style. The computer can fairly simply be instructed to process blocks of raw data, to count the words of one letter, two letters, three letters, and so on, or else to count the words found between full stops. Statistics of possible relevance concerning averages and distributions within each block of text can be prepared easily enough by the machine. But there are methodological difficulties. English spelling is so arbitrary that arguably it provides a poor guide to 'word length'. Only in a crude sense can 'bough' be regarded as longer than 'bow' and as long as 'unity'. However, if we want to compute word length in terms of the number of syllables, then the data will almost certainly require encoding. Again, sentence length presents difficulties. The punctuation we find in texts, especially in modernist works and in texts from the seventeenth century or earlier, often relates only erratically to the grammatical level. Furthermore, punctuation seems especially susceptible to the tamperings of compositors. It may be necessary to edit texts to secure consistency as a first stage. Canonical investigation based on word length or sentence length has fallen into a relative desuetude.

Word Frequencies

In each text or part of a text it is possible to count how often words occur. We may construct the average rate of occurrence of words within particular kinds of literature at particular periods, and relate the practice within a specified text or part of a text to that larger context. By looking at, for example, texts which we know to be by Kyd, together with a representative sample of plays by his contemporaries, we can compile a list of words which he used more frequently than others usually did, and a similar list of words that he used less frequently. A disputed text may then be processed to see whether its profile of positive and negative characteristics resembles Kyd's. Such operations call for sophisticated statistical manipulation[3]. However, the broad outline of such an investigative strategy is obvious enough. There is a related procedure that has often been adopted. In most kinds of normal English discourse, some common words or phrases are customarily available as synonyms. Thus, we find the pairs 'whilst' and 'while', 'among' and 'amongst', 'farther' and 'further'. A text may well be characterised by the ways in which such alternatives are drawn upon. Similarly, in certain restricted contexts, words which are not strictly synonymous are virtually interchangeable. For example, in Elizabethan drama, in similar contexts a playwright may select from quite a narrow list of oaths and ejaculations ('pah', 'tut', 'troth', and so on) in a way which may be sufficiently characteristic to distinguish his work from that of contemporaries. The possibilities for compiling all this sort of information by computer are obvious enough. As we have already described, the programs which generate concordances can also produce the sort of statistical information which these operations require[4]. It is a simple matter, once the data has been prepared, to generate word-lists of, say, Kyd and others, to see which words he favours or avoids and to see what is distinctive about the choices he makes between synonyms and similar alternatives.

Spelling

Sometimes it has been argued that, where alternative spellings are current, the choices made may be characteristic of a writer's practice and serve to differentiate it from that of others. Of course, there are problems with printed material, where compositorial activity has intervened, but it may still be worth considering. Choice between alternative spellings, like choice between synonyms, can be explored with the aid of the computer. The texts, which must not be normalised, are processed to produce wordlists, which are scanned to discern distinctions in orthography.

Grammatical Features

We have shown how some grammatical features of texts may be investigated by the computer-processing of information in which each word is encoded according to the grammatical category to which it belongs. Authorial attributions have sometimes been based on a grammatical analysis. For example, the grammatical categories which terminate sentences may seem differentiating features of style. If we have the encoded information for pertinent texts, it requires only a simple programming operation to perform the appropriate calculation. Much more complex approaches are possible. For example, the machine can be instructed to scan encoded data to identify the ways in which sequences of grammatical categories appear. Thus, perhaps one text may be characterised by a high incidence of the sequence 'noun-preposition-adjective-noun', another by the sequence 'noun-adjective-preposition-noun', and so on. Similarly, the actual variety of ways in which categories are combined may seem distinctive. For example, Milic has argued that

Swift's style is characterised, inter alia, by the large number of different grammatical sequences which are to be found within three-word segments of his texts[5].

THE RATIONALE OF PROBLEM RESOLUTION

We described above the structure of the problems which author identification and canonical investigation engage. We may consider now their resolution.

Where There is Reason to Doubt the Integrity of a Work

The work is sectioned, perhaps arbitrarily, perhaps according to divisions suggested by traditional theories of disintegrationist scholars. The characteristics of each section are determined, and the investigator decides whether it is probable that the variations that occur between sections could occur within the work of a single author, or whether some theory of multiple composition must be advanced.

Where the Work is Anonymous or Pseudonymous

The characteristics of a number of likely candidates for attribution are determined, as are the characteristics of the problematic text. The investigator then attempts to ascribe the text to an author on the basis of significant resemblances with and differences from the possible alternatives.

Where External Evidence Suggests the text is a Work of Either A or B

Texts certainly by A and texts certainly by B are examined for characteristics which clearly distinguish the practices of A from those of B. The same features are examined in the disputed text, and the text is located in the canon of whichever writer it more closely resembles.

Where External Evidence Suggests the Text is a Collaboration Between A and B

To discover who was responsible for the various parts of the work, texts known to be by A and texts known to be by B are examined for differentiating characteristics; the problematic text is sectioned, and the pertinent features of each section are examined to determine proximity to the known practices of A and B; and the probable author of each section is identified.

STATISTICAL CONSIDERATIONS

Investigations of the kind we have been considering, if they are to have a validity, need a scientific rigour. We have freely used in this chapter notions of probability and significance. These are not concepts to be invoked lightly. They relate to statistical propositions and procedures which assess, scrupulously, the levels of proximity and difference between texts and samples of texts. The 'proof' that a text is by a certain author or that it is not the work of one writer is probabilistic, and can only be arrived at through the application of appropriate statistical tests[6] and may remain elusive even then.

SUMMARY

This chapter discussed the nature of problems relating to author identification and canonical investigation and describes the strategies appropriate for their resolution and the role of the computer in such procedures. We stress the importance of a scientific rigour in interpretation of the statistical aspects of such investigations.

NOTES

1. The following are some of the more influential classical essays in author identification: Alvar Ellegard, *Who Was Junius?*, Stockholm, 1962, *A Statistical Method for Determining Authorship: The Junius Letters 1769–1772*, Gothenburg, 1962; F Mosteller and D L Wallace, *Inference and Disputed Authorship: The Federalist*, Reading, Mass, 1964; A Q Morton and James McLeman, *Paul, the Man and the Myth: A Study in the Authorship of Greek Prose*, London, 1966; Yehuda T Radday, *The Unity of Isaiah in the Light of Statistical Linquistics*, Hildesheim, 1973; A J P Kenny, *The Aristotelian Ethics*, Oxford, 1978. Most are

reviewed interestingly by Kenny, *Computation of Style*, pp. 6–10. See also E G Fogel, *Evidence for Authorship*, Ithaca, 1966, and Samuel Schoenbaum, *Internal Evidence and Elizabethan Dramatic Authorship*, London, 1966.

2. David J Lake, *The Canon of Thomas Middleton's Plays: Internal Evidence for the Major Problems of Authorship*, Cambridge, 1975, pp. 7–8.
3. See Alvar Ellegard, *A Statistical Method*, especially Chapter 2.
4. See above, Chapter 6.
5. Milic, op. cit., 204ff.
6. For some useful statistical guides, see footnote 4 to chapter eight. Pierre Guiraud, *Les Caractères Statistiques du Vocabulaire*, Paris, 1954, has been widely influential in canonical investigations.

10. FUTURE DEVELOPMENTS

Of course, our account of computer applications in literary studies is not exhaustive, nor did we intend it to be. Our examples have been drawn almost entirely from the study of English literature. Literatures in foreign languages sometimes pose different opportunities for computer applications, as well as occasional difficulties such as representing non-English alphabets to the machine for processing. Difficulties have very largely been resolved by the creators of concordance programs (though some Celtic languages which have initial mutations present special problems). Moreover, languages which are more nearly phonetic in their orthography can be analysed for rhyme, assonance and alliteration. Again, if a language is largely phonetic and its prosody observes strict rules, as, for example, in Welsh or in Latin, then the machine can produce analyses of a kind which are not possible for English-language texts. Also, current research in the field of stemmatics seems very promising, and computer-processing will no doubt play an increasingly important role in the establishment of relationships between alternative versions of the same text.

However, future revolutions in the role of the computer in the study of literature are likely to owe less to current research in literary applications than to major developments in microelectronics and computer science.

We have had much to say about preparing literary problems so that they can be presented to computers for resolution. Certainly, packages of programs and some programming languages make it easier than it used to be. But the concept of 'user-friendliness', of machines and systems which permit an easy dialogue with users and guide the presentation of their problems, can and will be extended very much further. Problems at the interface between user and systems mean that some technical expertise is still necessary, and this inhibits and discourages the literary computer-user. Five or ten years on, such obstacles may have almost disappeared.

In many institutions, computing facilities are a scarce resource, under pressure and competed for by many researchers. Though microcomputers offer an attractive alternative to mainframe systems, there are major limitations, especially in memory size, which restrict their usefulness. Current developments, however, are easing the problem, and it is reasonable to expect that the microelectronic revolution, continuing apace, will deliver a massive increase in the processing power available; indeed it is already doing 2.50, as the cost of memory is declining all the time. New microcomputers will be powerful enough for some literary applications which currently they cannot accommodate. New mainframes will be larger, quicker, cheaper, and based on new architectures. Moreover, communications technology, which already makes it possible to access large regional computer centres from remote sites, will make many more systems available, and may widely permit the linkage of systems, such as series of independent microcomputers, for the resolution of particular problems.

Major changes in the way we regard and process information will have far-reaching implications for the literary computer-user. We think of texts, the basis of all literary investigation, at the physical level as ink on paper. Whatever the future of the printed book and the traditional library, it is confidently predicted that, in the relatively near future, an increasing proportion of published material will be held on computer file. Data preparation has always been a major difficulty in the assembly of archives of literary texts on computer files. Optical scanners, which read printed pages directly into the machine, have in the past been very expensive and have required transcription of texts into a uniform and peculiar typeface. But modern books can be scanned quite cheaply by the latest systems, and no doubt, in time, even the vagaries and irregularities of the most badly printed incunablia will fall within the capacity of new technology. Holding thousands of texts, massive arrays of information, on file will produce a major breakthrough for literary computing. But let us consider the combination of that with current trends in communication technology, which will permit such files to be accessed remotely, from almost anywhere in the world, via optical fibres or communications satellites. Add to that the cheaper availability of computer hardware, providing a terminal in most offices, in most studies, perhaps even in most homes, and information science is transformed.

All this may revolutionise computer applications in literary studies. From a private room, say, in the United Kingdom, the researcher may formulate a problem, relating perhaps to a seventeenth-century text. Using a terminal, the researcher accesses a British Library catalogue to discover whether the text is on file and where it is stored. Perhaps it isn't on file in the UK. He or she dials instead a world catalogue, located maybe in California. This may answer that it is on file in Australia. Again using the terminal, the researcher contacts the Australian centre and requests that a copy of the file be switched by satellite to the computer centre the researcher has decided to use (selected from the hundreds or thousands accessible without leaving the terminal). The researcher establishes a dialogue with the computer, interactively defining the problem and how it is to be resolved, and the results are delivered in an appropriate form. All this, without leaving one's private room and with less technical expertise than it takes to operate a tape-recorder. Of course, if the researcher wants to scour the world's data bases for secondary material on the text—or if he or she just wants to read the text itself—that will be possible too. Prediction about technology is always a reckless undertaking, never more so than now but such a scenario for the use of the computer in the study of literature, fifteen or twenty years on, perhaps sooner, is not fantastic. Several times in this book we have discussed the difficulties of programming computers to function at the semantic level, to understand natural language. Even this challenge may be met. The problem lies in the fact that natural language can only be understood in its larger cultural context. But new generations of machines may have the storage capacity to have a mass of cultural information and the power to consult it intelligently and thus understand natural language in some of its richness, complexity and ambiguity.

GLOSSARY OF TERMS

The vocabulary of terms in common usage in computing and in its application to the study of literature is ever changing and a glossary needs to be frequently reviewed. In the interests of standardisation, many of the definitions we offer here are based on the third edition of the *Glossary of Computing Terms,* published by the British Computer Society.

ACM
   The US computing organisation, the Association of Computing Machinery.

ADDRESS
   The code, usually numerical, used to designate a specific storage cell; e.g., cell no. 17 has address 17.

ALGOL (ALGOL60, ALGOL68)
   A structured, high-level programming language. It appears as two distinct types, ALGOL60 and ALGOL68.

ALGORITHM
   A term used to describe a finite set of rules for solving a specific problem.

ALLITERATION
   The repetition of the same initial sound in words in close proximity to each other.

ANSI
   The American National Standards Institute.

ARITHMETICAL AND LOGIC UNIT
   The part of the central processing unit where arithmetical and logical operations are performed. Also called the arithmetic mill or unit.

ARCHIVE
   A store contained on some backing medium (e.g., disk or magnetic tape); it may be permanently on-line to the main store, or off-line, in the form of cassette, exchangeable disk or tape, etc..

ARTIFICIAL INTELLIGENCE (AI)
   A term used for the concept that computers can be used to assume some of the capabilities normally thought to be like

human intelligence, such as learning, decision taking, and self-correction.

ARTIFICIAL INTELLIGENCE LANGUAGES (AI languages)
Programming languages specifically designed for AI research and development on existing computing machines.

ASSEMBLER (also ASSEMBLY PROGRAM)
A program, usually provided by the computer manufacturer, to translate a program written in an assembly (symbolic-type) language to the machine code of the machine. Each assembly code instruction is changed into one machine-code instruction.

ASSEMBLY CODE (also ASSEMBLY LANGUAGE)
A symbolic language used for programming which must go through an assembler in order to be converted into the machine code required for operation on a computer.

ASSONANCE
A correspondence in sound between words which fall short of rhyme, as in the similarity of vowels in cane and rave, or the similarity of consonants in stick and luck.

AUTOCODE
A low-level programming language with a mathematical type of notation for the expression of the problems to be solved.

BATCH (or BATCH-PROCESSING)
A method of using the computer in which all the input (data and programs) are collected together ('batched') before processing begins.

BACKING STORE
A store for large amounts of information which has a slower access time than the immediate access store or main store of a computer.

BASIC
An aconym for Beginners' All-purpose Symbolic Instruction Code, a high-level, conversational programming language.

BINARY (NOTATION)
Usually refers to a system of using a base two and the digits 0 and 1 to represent information.

BUFFER STORE
An area of the store used temporarily to hold data being transmitted between a peripheral device and the central processor or another device to compensate for differences in their working speeds.

BUILT-IN REPERTOIRE
The computer manufacturer's set of machine code instructions which govern the basic operations of the computer.

CAI (CAL, CBL)
Stands for Computer Aided Instruction and refers to the use of the computer in education as an aid to the presentation of material. Also Computer Aided Learning and Computer Based Learning.

CAI LANGUAGES
Computer Aided Instruction languages. See course writing language.

CANONICAL INVESTIGATION
Investigation into works attributed to an author to establish which authentically belong to his canon.

CARDS (also PUNCHED CARDS)
An input medium in which information is represented by patterns of rectangular holes punched on cards.

CASSETTE
A magnetic tape cassette is a device for holding magnetic tape. The tape used may well be similar to that of a domestic tape recorder.

CELL (also called STORE or STORAGE LOCATION)
A basic unit within the computer store, capable of holding an item of information.

CENTRAL PROCESSING UNIT (CPU)
The main part of the computer, consisting of the immediate access store and the arithmetic and logic unit. It coordinates and controls the activities of all the other units and performs all the arithmetical and logical processes to be applied to the information.

CHARACTER
A letter, digit, or some other symbol that may be represented in a computer.

CHIP (SILICON CHIP, CHIP OF SEMICONDUCTOR MATERIAL)
The currently popular name for an integrated circuit which is a solid state microcircuit diffused into a single chip of silicon material.

COCOA
An acronym derived from COunt and COncordance generation

on Atlas, an early, machine-independent program to make concordances and indexes and to compute word-frequencies.

COLLOCATE

To place words together, as in a phrase or sentence.

COLLATION

Bringing together and comparison of different copies of a text in order to identify variants.

COMMAND LANGUAGE (also **JOB CONTROL LANGUAGE**)

A language designed for communicating with the operating and time sharing systems.

COMPILER (also **TRANSLATOR**)

A program which translates a high-level language program into the computer's own machine code or some other low-level language.

COMPUTER

A machine which, under the control of a stored program, automatically accepts and processes data and supplies the results of that processing.

COMPUTER ARCHITECTURE

The design of a computer and the way in which the hardware and software are constructed and made to interact to provide basic facilities and performance.

COMPUTER WORD

A collection of bits (1's and 0's) which are treated as a single unit by the central processor of the computer.

CONCORDANCE

A listing of, generally, every occurrence of words within a text, usually in alphabetical order, together with some of the context in which the words appear.

CONTEXT (also **CONTEXT** package)

A special purpose computer system for literary and linguistic applications.

CONTROL SIGNAL

A signal (set of electronic pulses) the function of which is to initiate controlling operations over the computer's devices.

CONTROL UNIT

Part of the central processing unit which supervises the execution of a program's instructions.

CONVERSATIONAL MODE

A method of operation in which the user appears to be in

continuous communication with the computer and receives immediate replies to his input messages.

CONVERSATIONAL PROGRAMMING LANGUAGES

Programming languages which allow the computer to be used in a conversational mode.

COURSE WRITING LANGUAGE

A programming language designed to assist the authors of course material to prepare computer programs.

CRT DISPLAY (or SCREEN)

Popularly used to refer to the display of information on a TV-type screen. CRT is an abbreviation for cathode ray tube.

DATA-BASE

A collection of structured information (data).

DATA-BASE MANAGEMENT SYSTEMS (DBMS or DATA-MANAGEMENT SYSTEMS)

The software used for the management and retrieval of the data stored in the data-base.

DATA-MANAGEMENT

The computer procedure by which information or records are input, stored, sorted and retrieved.

DATA PROCESSING

The complete operation of collecting information (data), processing it, and presenting the output results.

DATA STRUCTURE

The organised form in which items of information (data) are held in the computer, e.g., as a table or list.

DECISION SYMBOL

A flowchart symbol, usually a diamond box, to indicate that a decision has been made at that point.

DEVICE

A unit of the computer's hardware.

DIALOGUE ENTRY

Accessing the computer using a 'question and answer' technique, usually by way of an interactive terminal such as a VDU or teletype.

DIGITAL COMPUTER

A computer which operates with information represented in digital form.

DIGITAL FORM

The form of representing information through combinations of discrete pulses denoted by 1's and 0's.

DIGITIZED INFORMATION
See DIGITAL FORM.

DISK (also EXCHANGEABLE DISK)
A magnetic disk, which is a storage device consisting of a number of flat, circular plates, each coated with some magnetic material on both sides. Disks are usually fitted on a spindle and can rotate. Information is written to and read from a set of concentric circular tracks inscribed on the disk surface. Disks that can be removed are called 'exchangeable'.

DISK DRIVE (UNIT)
A mechanism to cause the disks to rotate on a spindle between 'heads' which allow the reading and writing of information.

DISK FILE
A file of information held on a magnetic disk.

DISPLAY SCREEN (or WRITER)
Usually refers to the cathode ray tube (CRT) or a visual display unit (VDU).

EASYWRITER
A wordprocessing package designed to run on Apple microcomputers. EasyWriter is a trademark of Apple Computers Ltd.

EDITING SYSTEM (EDITOR)
A program which allows information (i.e. programs, texts) to be modified (edited) by, for example, inserting or deleting characters.

EDITOR
See EDITING SYSTEM

ELECTRO-MECHANICAL (DEVICE)
A term used to indicate that a device is made up of both electrical/electronic and mechanical parts.

EXECUTION (of the program or instructions)
The part of the computer cycle at which the instruction is obeyed. In the case of a program of instructions, the point when the computer performs the intended functions.

FAMULUS
A machine-independent data-management system, particularly useful in bibliographical applications.

FILE
An organised collection of related records of information.

FILE MANAGEMENT
This refers to a systematic approach to the storage and retrieval of information stored in a file.

FLOPPY DISK
A lightweight, flexible magnetic disk which is used for storing information. It behaves as if rigid when rotated.

FLOW CHART
The graphical representation of the operations involved in the solution of a problem by the computer.

FLOW CHART SYMBOLS
The various graphical symbols used to represent the particular operations of a flow chart.

FLOW LINES
These indicate the sequence of operations or the flow of information in a flow chart.

FORTRAN
From FORmula TRANslation. The first symbolic programming language for a computer.

GENERAL PURPOSE COMPUTER
A computer that can handle a wide variety of tasks.

GRAPHICS SYSTEM
This allows the display of information in a graphical form on an output device. It usually incorporates a cathode ray tube for the display of both line drawing and text in black and white or colour. Light pens can be used to input or reposition the displayed information.

GRAPH PLOTTER
An output device which draws lines on paper.

HARD COPIER
An output device to print information on paper.

HARDWARE
Physical units making up a computer system, such as electrical, electronic, magnetic, mechanical devices, as opposed to the computer programs (the software).

HIGH LEVEL LANGUAGE (also HL PROGRAMMING LANGUAGE)
A problem-oriented language in which instructions may be equivalent to several basic machine-code instructions and which may be used on different computers with an appropriate compiler.

HIGH QUALITY PRINTER
A sophisticated printer which has the facility to produce high quality print forms.

HIGH-SPEED DIGITAL COMPUTER
See COMPUTER

HOMOGRAPH DISTINCTION
The separation, in the process of compiling a concordance, of different words which share the same spelling.

INFORMATION PATTERNS
The representation of information in a coded form, usually as sets of 1's and 0's, which can be handled by the hardware and software of a computer.

INFORMATION RETRIEVAL LANGUAGES
Programming languages which are especially designed to allow for the efficient retrieval (and partial manipulation) of stored information.

INPUT
Information to be taken into a computer.

INPUT DEVICE
The equipment used for taking information into a computer.

INSTRUCTION
A statement which tells the computer which operation to perform.

INTERACTIVE COMPUTING
The way of operating a system which allows the terminal user and the computer to communicate with each other.

INTERFACE
A shared boundary, e.g., between two devices, subsystems, or the user and the system. The last is termed the man-machine interface.

INTERPRET (or DECODE)
The process of translating and executing the statements of a program one at a time.

INTERPRETER
A computer program which translates and executes a program one statement at a time.

JOB-CONTROL LANGUAGE (JCL)
A language to recognize a computer program and its data, called a 'job', and to describe its demands to the operating system.

JUSTIFICATION
In printing or wordprocessing, the process of adjusting the spacing between letters or words to fit exactly the length of the line of type from margin to margin.

KEYBOARD
A device, such as a teletype, for encoding information by the depression of keys.

KEY-TO-DISK
  The process of transferring information from a keyboard and writing it directly to a magnetic disk.

KEY-WORD
  In a concordance, the indexed word around which context is arranged.

LEMMATISATION
  The process of grouping together words that are inflected or variant forms of the same word. Thus, a concordance may list together am, is, are, was, were, etc. under the verb be.

LEXIS
  The aspect of language concerned with vocabulary.

LIST PROCESSING
  The processing of information which has been structured into items and stored in the computer as a 'list'.

LIST PROCESSING LANGUAGES
  Programming languages especially developed for list processing, such as LISP 1.5, MLISP.

LITERARY COMPUTER
  A computer used for literary application. Usually contains software specially developed for the user who is involved in literary studies.

LOW LEVEL LANGUAGE
  A language in which each program instruction has a single corresponding machine code instruction.

MACHINE
  The 'computing machine'. See computer.

MACHINE CODE (also MACHINE CODE LANGUAGES)
  The manufacturer's set of basic fixed instructions that govern the operations of the machine and that are available to the user for programming.

MAN-MACHINE INTERFACE
  See INTERFACE

MACHINE-INDEPENDENT
  Not dependent on a particular computer: e.g., a 'machine-independent' programming language.

MACHINE-ORIENTATED LANGUAGE
  A programming language that is based as closely as possible on the machine code of the target computer.

MACHINE RANGE
A family or series of computing machines produced by a manufacturer, e.g. the ICL 2900 series.

MACHINE REPERTOIRE
See built-in repertoire.

MACHINE'S FUNCTIONS
See built-in repertoire.

MAGNETIC TAPES (also EXCHANGEABLE TAPES)
A storage medium which consists of a flexible plastic tape covered with magnetic material on one side and wound on a reel or spool.

MAINFRAME COMPUTER
A computer with a variety of peripheral devices, a large amount of backing store and a fast central processing unit (CPU). The term is used in distinction from smaller computers.

MAIN STORE
The principal fast or immediate access store (IAS) of a machine. Also known as the primary store or memory.

MATCHING
A technique to compare items of information, such as a set of symbols or characters grouped to form a string or a pattern.

MATHEMATICAL PACKAGES
Programs developed for mathematical applications.

MEMORY (or STORE)
The part of the computer where information is held.

MICROCOMPUTER
A computer, the central processing unit (CPU) of which is a microprocessor.

MICROPROCESSOR
A single chip which performs the function of a central processing unit (CPU).

MICROPROCESSOR UNIT (MPU)
A device which contains at least one microprocessor.

MINICOMPUTER
A computer which, in its size, speed and capabilities, falls between a mainframe computer and a microcomputer. Usually refers to a range of machines which is less well equipped than the mainframe machines.

MONITOR
The control or supervisor program which schedules the use of the hardware and software required by a program that is being executed, which is thus 'monitored'.

MOVING SURFACE MEMORIES (or DEVICES)
Devices which allow information to be stored in particular regions of some magnetic material, such as plastic tape or disk, which forms a moving surface.

MULTIPURPOSE LANGUAGE
Programming languages which are sufficiently versatile to be used in a variety of applications.

NEOLOGISM
A newly coined word or the practice of using such words.

NETWORK (or COMPUTER NETWORK)
A set of computer systems, usually geographically apart, which are linked so that they can communicate with each other and, if desired, share facilities.

NON-NUMERICAL INFORMATION
Information which does not consist solely of numerals and which may contain alphabetical characters or other symbols.

NON-NUMERICAL LANGUAGES
A class of programming languages specifically designed to solve problems involving non-numerical information.

NORMALISATION
Often, especially in works from the early modern and earlier periods, variations in spelling and punctuation occur within a text. In preparing these texts for publication or for concordance generation, it is sometimes appropriate to reduce each set of variations to conform to one of the alternatives. This process is termed normalisation.

OCP
The Oxford Concordance Program, a machine independent program to make concordances and indexes and to compute word-frequencies.

OPERATIONS
Defined actions of the computer.

OPERATING SYSTEMS
Specially written programs to control the running of a computer system with the minimum of human operator intervention.

OPTICAL READER (also known as OCR, OPTICAL SCANNER or OPTICAL CHARACTER RECOGNITION DEVICE)
An input device which recognises characters by light-sensing methods and reads them into the computer.

ORTHOGRAPHICAL IRREGULARITY
Irregularity in spelling within a text, occasioned when the same

words recur in alternative spellings, a common characteristic of works in English from the early modern period and earlier. See NORMALISATION.

OUTPUT

The results produced by a computer when it transfers information from internal storage into an external readable form or into a form suitable for re-processing by the computer.

PACKAGES

See SOFTWARE PACKAGES.

PASCAL

A structured ALGOL-like, high-level programming language.

PATTERN

See INFORMATION PATTERN.

PERSONAL COMPUTER (or HOME COMPUTER)

A microcomputer that has been 'packaged' for personal use.

PHONOLOGY

The aspect of linguistics which is concerned with the study of languages as sound systems.

PRINTED CIRCUIT BOARD (or PCB)

A rigid plastic sheet which incorporates the principal microelectronic components in a microcomputer.

PRIMARY STORE (or PRIMARY MEMORY)

See MAIN STORE.

PRINTER (also LINEPRINTER)

An output device which converts information into a printed form.

PROBLEM ORIENTATED LANGUAGE

A programming language designed for the convenient expression of a problem.

PROBLEM-SOLVING INTERFACE

An interface designed specifically to help the user communicate his problem to the computer and to assist in obtaining its solution.

PROBLEM-SOLVING LANGUAGES

Programming languages designed for the convenient expression of a problem.

PROCEDURE

A set of program instructions designed to perform a specific task, but which is not a complete program.

PROCESSOR

The name of any hardware device capable of carrying out operations on data. Also, in software, a computer program which includes operations such as compiling, assembling and related tasks: e.g., a FORTRAN processor.

PROCESSING OPERATION SYMBOL

A flowchart symbol (usually a rectangular box) to indicate that an arithmetical or logical operation has to be carried out.

PROCESSING UNIT

A part of the computer that contains at least one processor, such as a CPU or MPU.

PROGRAM

A complete set of instructions arranged in such a way as to solve a given problem by computer.

PROGRAM DATA

Information required by a program during its execution.

PROGRAMMER

The computer user who prepares instructions to solve a problem on a computer.

PROGRAMMING LANGUAGE

An artificial language designed to allow the user to communicate problems to the machine in a precise and intelligible fashion.

PROM

Programmable Read-Only Memory is a memory storage of the ROM type, where a program may be written after manufacture by the user but which can be changed from time to time.

PROTEXT PACKAGE

From PROcessing TEXT: a set of programs for use in literary and linguistic applications.

QUERY LANGUAGES

Languages specially designed to allow the user to access stored information.

QUESTION/ANSWERING SYSTEMS

Computers which have been programmed to conduct a dialogue of questions and answers with the user.

RANDOM ACCESS (also DIRECT ACCESS)

The process of storing or retrieving information without the need for any other information to be read first.

RANDOM-ACCESS MEMORY (or RAM)

A memory which may be read from and written to by the user. Usually made on a chip which is incorporated into a microcomputer system.

READERS (also TAPE READERS and CARD READERS)

Devices designed to read information, e.g., from tape or cards.

READ-ONLY MEMORY (ROM)
A memory which cannot have any information written to it by the user. The information in the ROM is fixed during manufacture.

ROUTINES (also SUBROUTINES)
See PROCEDURE.

RUNTIME ERRORS
Errors which occur during the execution of a program.

SCREEN
See CRT display and VDU display.

SEMANTIC LEVEL
The aspect of language which is concerned with meaning.

SIMULATION (also COMPUTER SIMULATION)
A procedure to represent certain features of the behaviour of one system by the behaviour of another, e.g., by using a model of the system and processing it on the computer.

SNOBOL (also SNOBOL4)
From String Oriented Symbolic Language: a programming language designed for the processing of non-numeric information.

SOFTWARE
The set of computer programs concerned with the operation and application of a computer system.

SOFTWARE PACKAGE
A fully documented program or set of programs designed for particular tasks, e.g., FAMULUS, SPSS, OCP.

SPEECH INPUT DEVICE
A device to input the spoken word into a computer system.

SPSS
The Statistical Package for the Social Sciences, a machine-independent suite of programs for statistical analysis. A trademark of SPSS Inc.

STATISTICAL PACKAGES
Programs for statistical applications, e.g., SPSS.

STORAGE REGISTERS
Special store for information which has a particular function during arithmetical and logical operations: often used for specific purposes, e.g., for control purposes or to store results.

STORE (or MEMORY)
The part of a computer system where information (data, programs, etc.) are held.

STORE HIERARCHY
The levels of storage, as, for example, in the relationship of main store, backing store, buffer store.

STORE LOCATIONS (or MEMORY LOCATIONS)
Identifiable units for the storage of information.

STORED PROGRAM
A set of instructions in the memory which specify the operations to be performed by the computer.

STRING
A sequence of characters or digits.

STRING PROCESSING
The process of manipulating strings.

STRING PROCESSING LANGUAGES
Languages used for string processing.

STRUCTURED LANGUAGE
A language which is not restricted to single individual statements, but which allows a program to be written which reflects not only the structure of the problem and the data involved but also any constraints imposed on the solution. Characterised by reducing the need for branching instructions.

SUITES OF PROGRAMS
See SOFTWARE PACKAGE.

SYNTAX
The aspect of language concerned with the grammatical arrangement of words.

TAPES (also PUNCHED TAPES)
Paper tape which has been perforated with a pattern of holes to represent information. 'Tape' can also refer to magnetic tape.

TARGET MACHINE
A computing machine on which one expects to solve a problem.

TELETYPE
The commercial name for a particular make of tele-typewriter which is an input/output device consisting of a keyboard and a typewriter-like printer.

TERMINAL (also INTERACTIVE TERMINAL)
A term used to describe any input/output device which is used to communicate with the computer.

TEXT EDITING LANGUAGES
Programming languages developed especially for the editing of texts.

TEXT FORMAT
    The layout of a text with particular relation to its input, storage
    and output.
TIME SHARING SYSTEM
    An arrangement which allows multi-access to a computer system,
    each user being allowed, in turn, a slice of the system's resources,
    although each appears to have continuous usage.
TRACK (of a DISK)
    See DISK.
TRANSISTOR
    A small solid-state semiconductor (crystal) which can operate as
    an amplifier or as a fast switching device: usually made of silicon
    or germanium.
TRANSLATOR
    See COMPILER.
USER
    Anyone who makes use of a computer system.
USER-FRIENDLY SYSTEM
    A computer designed so that the interface between the user and
    the machine is markedly sympathetic to the user and requires him
    to know little about either the hardware or software that is being
    used.
VARIANT
    Points of difference between different copies of a text.
VDU SCREENS
    See VISUAL DISPLAY UNIT.
VISUAL DISPLAY UNIT (VDU)
    A terminal device which has a keyboard and incorporates a
    cathode ray tube as the screen on which information can be
    displayed.
'VON NEUMANN' COMPUTER
    A computer designed according to the ideas of the scientist and
    mathematician John von Neumann, who introduced the concept
    of stored program control.
WORD FREQUENCY
    The frequency with which different words occur within a text or
    texts; sometimes used for the relationship between the number of
    different words and the total number of words within a text or
    texts.
WORDPROCESSING
    A computer procedure which allows the input, storage and

alteration of texts and the output of these texts in a form analogous to the product of a typewriter or in a camera-ready form appropriate for printed reproduction.

WORDSTAR

A wordprocessing package, designed to run on microcomputers using the CP/M operating system. WordStar is a trademark of the MicroPro International Corporation and CP/M is a trademark of Digital Research.